T0218234

Lecture Notes in Computer Science

Lecture Notes in Computer Science

Edited by G. Goos and J. Hartmanis

295

R. Dierstein D. Müller-Wichards
H.-M. Wacker (Eds.)

Parallel Computing in Science and Engineering

4th International DFVLR Seminar
on Foundations of Engineering Sciences
Bonn, Federal Republic of Germany, June 25/26, 1987

Springer-Verlag
Berlin Heidelberg New York London Paris Tokyo

 DFVLR Seminar Series

organized by
Deutsche Forschungs- und Versuchsanstalt
für Luft- und Raumfahrt e.V. (DFVLR)
Linder Höhe, Postfach 90 60 58, D-5000 Köln 90

The text was produced for camera-ready reproduction using
the DFVLR word processing system GML.

CR Subject Classification (1987): C.1.1−2, C.2.1, C.4, D.1.3, D.2.6, G.1.8

ISBN 3-540-18923-8 Springer-Verlag Berlin Heidelberg New York
ISBN 0-387-18923-8 Springer-Verlag New York Berlin Heidelberg

© Springer-Verlag Berlin Heidelberg 1988
Printed in Germany

Printing and binding: Druckhaus Beltz, Hemsbach/Bergstr.
2145/3140-543210

Contents

Preface

The 1987 seminar on "Parallel Computing in Science and Engineering" was the fourth in a series of annual conferences on "Foundations of Engineering Sciences" hosted by the German Aerospace Research Establishment (DFVLR). It was held in honor of

Professor Dr.rer.nat.
Hermann L. Jordan

past president of the DFVLR. Three previous conferences covered highlights of

1984   Nonlinear Dynamics of Transcritical Flow[1]

1985   Uncertainty and Control[2]

1986   Artificial Intelligence and Man-Machine Systems[3]

Parallel Computing is required in a number of significant applications in science and engineering where classical sequential processing cannot supply - at least at present - the computing power in demand. In order to take advantage of the performance potential of parallel computer architectures it is, however, essential to map the application efficiently onto the architecture in question.

To this end a considerable effort with respect to the investigation of new algorithms and the development of adequate programming environments is necessary. The seminar is a survey on the state of the art in several important fields of Parallel Computing and discusses a number of issues in dealing with various applications.

---

[1] Lecture Notes in Engineering. Vol.13, Springer 1984
[2] Lecture Notes in Control and Information Sciences, Vol.70, Springer 1985
[3] Lecture Notes in Control and Information Sciences, Vol.80, Springer 1986

Part I - Introduction to the Seminar

Introduction to the Seminar

Hans-Martin Wacker

German Aerospace Research Establishment (DFVLR)
Post Weßling Obb.
8031 Oberpfaffenhofen

1. The Development of Scientific and Engineering Data Processing during the Last Thirty Years

It is now more than 30 years ago that a rapid and dramatic development began in the field of scientific and engineering data processing. Within just a few years the performance of computing machines had improved from several hundred FLOPS[1] to 20,000 FLOPS (20 KFLOPS). This new performance category had been opened up by the announcement of computers like e.g. the IBM 709.

At the same time FORTRAN, a new *high level* programming language had been developed. Its availability helped to considerably reduce the programming effort due to the use of assembly language.

As an almost natural consequence of the advent of powerful computers and high-level language compilers, those "computation centres" disappeared within a very short period of time where squadrons of mathematicians equipped with mechanical or electro-mechanical desk-top computing machines worked on the numerical solution of scientific and engineering problems.

The introduction of electronic computers which, properly programmed, were able to solve numerical problems much faster and far more efficient took only a few years. In a few seconds a single IBM 709 computer could execute more operations than one of the former "computation bureaus" in one day. Mathematical problems could not only be solved much faster than ever before, but it also became feasible to attack and solve entirely new problem classes considered beyond reach so far. A number of revolutionary results were achieved which in turn stimulated the development of still more powerful machines. Within a very short timeframe a sequence of ever-faster computers was announced. In 1970 two new computers, the CDC7600 and the IBM/360-195, capable of a peak performance of more than 3 MFLOPS entered the market. Hence within in a period of barely ten years computer performance again had increased by a factor of ten.

Although there was no sign of a slow down in technological innovation, after 1970 a period of stagnation - in some cases even regression - occurred, at least as far as the further increase of performance was concerned. Table 1 shows the performance of the IBM top mainframes between 1957 and 1985. In a sense Table 1 reflects the significance of engineering and scientific data processing in IBM's development policy.

---

[1]   FLOPS   Floating Point Operations per Second
[2]   Per Central Processing Unit (CPU)

Year	Type	Perfomance
1957	709	0.02 MFLOPS
1961	7090	0.05 MFLOPS
1967	360 - 75	0.3 MFLOPS
1968	360 - 91	2 MFLOPS
1970	360 - 195	3 MFLOPS
1974	370 - 168	0.8 MFLOPS
1978	3033	2 MFLOPS
1981	3081 D	2 MFLOPS[2]
1982	3081 K	2.5 MFLOPS[2]
1985	3090	7.5 MFLOPS[2]

Table 1.   Performance of Computers

The reason for a certain stagnation during the 70's was that a further increase in computer performance would not have been accompanied by a corresponding broadening of the spectrum of those applications where the performance potential of these sophisticated machines could have been utilized in an economically feasible way.

Consequently IBM withdrew its scientifically oriented system /360-195 from the market, substituting it by the less complicated /370-168 which was almost as efficient for non-numerical applications. For computationally intensive problems, however, only one fourth of the performance of its predecessor was attained.

Control Data Corporation had established her position as a leading vendor in the engineering-scientific market market through the CDC6000-series. CDC was not affected by the changing market as much as was IBM, although the performance of the CDC7600 was unsurpassed by any of its successors for quite a long time. Until 1986 CDC's leading edge models all had approximately the same peak performance as the 7600.

From 1970 on most manufacturers turned away from developing *Supercomputers* and concentrated on the design of affordable hardware. Computer systems became cheaper and the size of main memories increased in accordance with the ever advancing computer technology.

It was not until the late '70s that engineering-scientific computing received new impulses, primarily induced by the introduction of the first succesfull vector computer, the Cray 1. The perfomance of this machine was an order of magnitude higher than that of any other system commercially available. One of the major reasons for its success was the application field of seismic exploration. The classical method of oil prospection by drilling probes is exceedingly costly. Simulation and numerical evaluation of geological data on a vector computer required enormous computer performance but turned out to be far more economical than the trial and error method.

In 1965 a "large scale" mainframe like the IBM 7090 had still been considered big enough to provide sufficient computing power for a whole country like the Federal Republic of Germany. A few years later single companies, single research institutions or universities each were running computing centres with more powerful CPUs. Today, any single user may have at his fingertips a PC five times as fast as the IBM 7090 twenty years ago.

2. Considerations on Solving Multidimensional Boundary Value Problems

Already in the '60 s the fact had been realized that the set of solvable problems had initially experienced a dramatic growth with increasing computing power, but also a slow down later on. In the well known book on numerical methods by Forsythe and Wasow [FoW60] the authors published Table 2 where this relationship was demonstrated.

In this table the computational requirements for the solutions of the following class of elliptical boundary value problems in S dimensions were listed:

$$L U = q \text{ in } G$$

$$U = \varphi \text{ in } \partial G$$

If the step size is

$$h = \frac{1}{N},$$

the number of grid points will be proportional N^s.

The IBM 709 was taken as a reference machine and SOR was used as a solution method for the difference equations.

	N = 10	N = 100	N = 1000
S = 1	1 sec	1 min	1 h
S = 2	3 sec	1 h	6 weeks
S = 3	1 min	1 week	100 yrs

Table 2. Estimated computing time for the solution of the elliptic boundary value problem in S Dimensions on an IBM 709 in 1960 *(Forsythe-Wasow 1960)*

Obviously in 1960 the two-dimensional case with $N = 1000$ and the two three-dimensional cases with $N = 100$ and $N = 1000$ were unsolvable, i.e. beyond the range of an IBM 709. Table 3 and Table 4 represent the situation ten years later and today respectively.

In spite of a hundred fold increase w.r.t. computer performance of the IBM 360-91 over the IBM 709 the gridsize corresponding to $N = 100$ for the 3-D-case remained out of reach. Today - even if a Cray-2 is used - the problem for $S = 3$ and $N = 1000$ can still not be solved employing the methods of the '60 s. Since that time, however, considerably more efficient numerical schemes have been developed that can be utilized to solve these elliptic boundary value problems.

	N = 10	N = 100	N = 1000
s = 1	0,01 sec	0,6 sec	40 sec
s = 2	0,03 sec	40 sec	10 hr
s = 3	0,6 sec	2 hr	5 weeks

Table 3.   1968: Estimated computing time for the elliptic boundary value problem on an IBM 360-91 (2 MFLOPS)

	N = 10	N = 100	N = 1000
s = 1	0,0002 sec	0,01 sec	1 sec
s = 2	0,0006 sec	0,7 sec	12 min
s = 3	0,01 sec	2 min	15 hr

Table 4.   1987: Estimated computing time for the elliptic boundary value problem on a Cray 2 (100 MFLOPS)

In the meantime attention had turned towards a different class of problems. The solution of linear elliptic equations has become less important for many practical problems. An adequate description of fluid flow, the behaviour of a plasma, or the simulation of a car-crash lead to highly nonlinear partial differential equations whose solution is extremely cumbersome and computationally intensive. In fact, as far as the solution of these problems is concerned, we have barely attained a position comparable to that described by *Forsythe* and *Wasow* for elliptic partial differential equations.

As an illustration, weather-forecasting-models as they are executed on today's most powerful vector computers like e.g. the Cray-XMP use a stepsize of approximately 100 km. With this stepsize, however, it is impossible to model a geographical formation like the Alps which play a decisive role for the weather in central Europe, in an appropriate way. In order to obtain an acceptable approximation for the Alps the stepsize has to be reduced to 10 km - or even less.

The amount of computer operations for these three-dimensional problems, however, is proportional to $1/h^4$. This would mean that the computational requirements would increase by a factor of 10^4, i.e. a computer would be necessary with 10^4-times the performance of a Cray-XMP, at least, if we stick to uniform grid spacing.

In other words, a 10-fold increase of computational power would bring about only a minor improvement of the result. The resolution would only grow by a factor of $\sqrt[4]{10}$.

3. Limits of Performance Increase by Vectorization - Amdahl's Law -

3.1 Vectorization of Programs

Usually, programs for the solution of problems in mathematical physics or engineering sciences consist of a mix of vectorizable instructions and scalar instructions as well. There is a direct relation between the proportion of vectorizable instructions within a program and the increase in performance obtainable by using vector computers. Normally, it is the part of the program for the mathematical solution of the problem which contains the largest proportion of possibly vectorizable instructions.

Certain parts of a program tend to be not vectorizable at all or only to a very limited degree. Among these are the transformation of the engineering or physical problem into a mathematical description, the back transformation of the mathematical solution into physical results, and the graphical representation. Other parts of programs which are barely or not at all vectorizable are the handling of boundary conditions and above all the data management in the case of insufficient memory (overlay structures).

The fact that only a certain portion of an application can be processed in a high performance mode had been pointed out by Gene Amdahl, orginally in the context of multiprocessors. Thus the basic law describing the relation between the performance of a vector computer and the degree of vectorization is called *Amdahl's Law*.

$$P = P_S \, \frac{\beta}{\alpha + \beta(1 - \alpha)}$$

The elements of this function are

P overall performance

P_s scalar performance

P_v vector performance

α degree of vectorization

$\beta = \dfrac{P_v}{P_s}$ vector-scalar performance ratio

Figure 1 shows the performance as a function of the degree of vectorization for various vector computers following Amdahl's Law. Obviously vector computers with a very high vector scaler performance ratio will only achieve a high overall performance for programs with a high the degree of vectorization. For programs which are vectorizable only to a lesser extent the total performance decreases drastically for this type of computers.

3.2 Programming and Vector Length as Influence Factors

We have seen in the previous section that a high degree of vectorization is a prerequisit for the economical use of vector computers. This however, is only a *necessary* condition which by now means is already a guarantee of good performance. With the results of his benchmarks published in [Don871] and

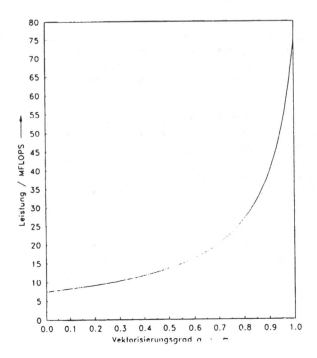

Figure 1. Performance of vector computers as a function of the degree of vectorization

[Don872] J.J. Dongarra showed how carefully the relation between vectorization and performance has to be evaluated. He investigated the performance of a large number of different vector processors comparing the results for two different sets of programs. The first one was a program of the LINPACK library for the solution of a dense system of linear equations of the order of 100. In the second test the program was optimized and especially tailored to the use of the various types of vector computers solving the same linear problem for the order 300.

The results of his benchmarks showed considerable differences in performance. Dongarra mentions two reasons for these differences despite an almost complete vectorization in both cases. The differences are primarily due to the different programming techniques used in the two programs. The influence of larger vectors in the second case, however, appear to be only of second order. The different behaviour demonstrated in Figure 2 and Figure 3 where the performance of four vector computers dependent on the degree of vectorization is shown.

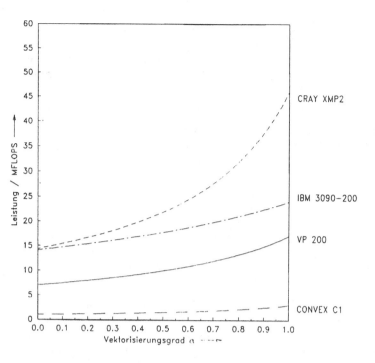

Figure 2. Benchmark results using a LINPACK program kernel *(J. J. Dongarra)*

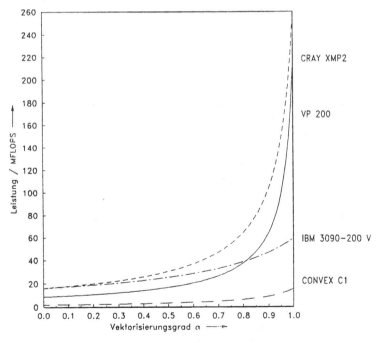

Figure 3. Benchmark results using an especially tailored program kernel *(J. J. Dongarra)*

A comparison of the graphs in both figures induces an interesting conclusion with respect to the economical use of vector features. Considering the costs of the systems and looking at the more expensive vector computers these systems are superior in performance only in those cases, where the overall load has a degree of vectorization of at least 80%, and where the programming is especially tailored with respect to the hardware structure of the vector computer under consideration.

In all other cases it is more economical to use a universal computer with an additional vector feature or a mini vector computer instead.

4. Future Trends of High Performance Processing

In most cases the *scalar part* of a vector computer is quite the same hardware as the central processing unit (CPU) of a high performance universal computer. Fujitsu e.g. uses for its vector computers VPxxx a scalar unit which is only a slightly modified version of the CPU of its model M380.

This means that for the technical development to be expected during the next years the scalar performance of vector computers will increase only to the same degree as the overall performance of universal computers. Despite a considerable increase in performance of switching elements and circuits due to the improvement of basic circuit technology the overall performance increase of CPU's will be much more conservative. The problem of integrating a large number of components into one single central unit will make the CPU overall performance highly dependent on the communication necessary for the interaction between the various components. Hence, during the next five years scalar performance might be expected to increase from a maximum of now 10 MFLOPS to perhaps only 20 MFLOPS.

This sort of physical limitation does not exist for the *vector performance*. Vector performance may be enhanced almost arbitrarily if only expenses for the vector processors and for the efficiency of the working storage are sufficently high. In principle it should be possible to develop in the forseeable future vector processors having a vector performance of several thousand MFLOPS. Taking into account the expected scalar performance of only 20 MFLOPS mentioned before, single processor systems with a vector/scalar performance gap of this kind may be reasonably useful only for a rather narrow spectrum of specialized problems.

To achieve useful utilization of such a *high vector performance single processor system* the degree of vectorization for the underlying programs must be very high (>95%). Therefore vector computers of this type are especially useful for linear problems, e.g. elliptic boundary value problems, since for this type of problems the degree of vectorization converges to 100% with decreasing stepsize provided the working storage is sufficiently large.

The situation with many of today's practical problems is somewhat different. Not only are many problems highly nonlinear but also the geometry is becoming increasingly complicated, leading typically to domain decompositions in order to avoid geometric singularities. Growth in gridsize is primarily spent on gaining sufficient resolution for more complex problems rather than on convergence for a fixed problem. Consequently the obtainable degree of vectorization is limited exactly for those problems requiring the highest computer performance. Hence, the obtainable overall performance of a technically feasible vector computer for these cases is limited as well.

Most manufacturers of supercomputers have thus tried in the meantime to over-come the performance limitations of vector computers by developing tightly cou-pled *multiprocessor systems*. Only the use of a larger number of CPUs will push the scalar performance beyond the physical limits of single processors. Multiple processor systems may be a possibility to economically use vector processors for a class of programs with a degree of vectorization below 90%.

A quite different approach to the problem of performance increase has become feasable with the rapid development of more and more powerful micro processors. Integration of a CPU into a single chip avoids the communication delays between components, due to the extremely small geometrical dimensions in VLSI circuits. The considerable increase in computing power and efficiency of micro processors is therefore quite understandable.

VSLI techniques and the development of powerful and cheap micro processors are the foundations on which the old idea of *massively parallel computer architec-tures* can be put into realisation. Investigations towards parallelism have received strong new impulses and created a wide spread interest during the last few years. Unfortunately parallel multiprocessor systems create a new class of rather difficult problems, especially those of coordination, communication and interaction of many independently operating processors. Many investigations in research insti-tutions and industry concentrate on solutions to these problems, which may help to make parallel computation a feasible step towards higher computer perform-ance.

The contributions of the 1987 DFVLR Seminar are part of this effort and interest.

5. References

[Don871] Dongarra, J.J.
Supercomputer Performance Considerations
Proceedings of the Tutorial *Supercomputer '87*, Mannheim, June 12-13, 1987

[Don872] Dongarra, J.J.
Reconstructing Supercomputer Algorithms
Proceedings of the Seminar *Supercomputer '87*, Mannheim, June 12-13, 1987

[FoW60] Forsythe, G.E.; Wasow, W.R.
Finite Difference Methods for Partial Differential Equations
Wiley & Sons, 1960

Classification and Evaluation of Parallel Computer Systems

Roger W. Hockney

Computer Science Department
Reading University
Reading, Berks. UK, RG6 2AX

Abstract

A classification is presented of both vector (i.e. SIMD) and parallel (i.e. MIMD) computers, and two simple benchmarks are defined to assess performance. Results are presented for the Cray X-MP vector computer, and the LCAP parallel system of ten FPS-164 computers.

1. Introduction

In the last few years a wide variety of multiprocessor (or replicated) computer systems have appeared on the market, and as experimental systems in universities and research laboratories. Such systems range from small numbers (generally 2 to 8) of high-speed expensive processors to relatively large numbers (of the order 10 to 100) of much cheaper and inevitably much slower processors. Because the processors work independently and simultaneously, these systems are generally referred to as "Parallel Computer Systems". in contrast to the traditional single-processor serial von-Neumann architecture. The term parallel can also be applied to the overlapping or simultaneous operation of different stages of an arithmetic operation on successive elements of a vector, using the principle of pipelining. This is used in the highly successful vector computers, such as the Cray X-MP, CDC Cyber 205, Fujitsu VP and IBM 3090-VF. In this paper we present a classification of both types of parallel computer, and follow this with a description of two simple benchmarks which can be used to assess some aspects of the performance of these computers.

2. Classification

From the programming point of view, the most important distinction to be made in a classification is that introduced by Flynn [1], namely the number of instruction streams available. The first generation of vector computers (e.g. Cray-1, CDC Cyber 205, Fujitsu-VP) were controlled by a single instruction stream but introduced vector instructions to speed-up the execution of the same operation between successive elements of two vectors (e.g. element-by-element vector add instruction). The vector instructions were implemented by pipelining, that is to say by streaming successive elements of the vectors through a single multi-stage (typically six stage) pipeline, one clock period per stage. Thus, in this case, although each element spends six clock periods in the pipeline, one operation is completed every clock period when the pipeline is full (an asymptotic speed-up of six-fold). The speed-up takes place because different stages of an arithmetic operation on different elements are overlapped in time. A single vector instruction controls the manipulation of many data items, and in Flynn's terminology these are SIMD (Single-Instruction/Multiple-Data stream) computers.

An alternative implementation of SIMD computing is to assign each element of the vector operation to a different processing element (PE) of an array. Such a processor array simultaneously computes all elements of the result vector, still under the control of a single stream of instructions. These two types of SIMD computer are distinguished by naming them respectively "Pipelined SIMD" and "Replicated SIMD". The former have proved most successful for high-speed general-purpose Fortran scientific computation, but the latter is attractive for certain applications, particularly in image processing. The ICL DAP and Goodyear MPP are examples of Replicated SIMD computers.

If the computer system provides the programmer with more than one instruction stream to apply to the solution of a single problem, it is classified by Flynn as MIMD (Multiple-Instruction/Multiple-Data stream). The vast number of different MIMD computers that have appeared in the last few years, form a confusing menagerie of computer designs. In order to put these into some kind of order, we show in Figs 1, 2 and 3 a possible taxonomy. In the description below we give only the acronyms of the different computers, but a more detailed description of their individual architectures can be found in Hockney [2].

Multiple instruction streams may be processed either by time-sharing a single sophisticated pipelined instruction processing unit, or by providing separate (and necessarily much simpler) instruction processing hardware for each stream. In the latter case, this hardware is typically a commercially available microprocessor. The first alternative is described as Pipelined MIMD, and was used within a single process execution module of the Denelcor HEP, the first commercially available MIMD computer. MIMD systems using the second alternative naturally divide into those with a separate and identifiable switch (Switched MIMD), and those in which the processing elements are connected in a recognisable and often extensive network (MIMD Networks). In the former all connections between the processors are made via the switch which is usually quite complex and a major part of the design. In the latter, individual processors may only communicate directly with their neighbours in the network, and long distance communication across the network may require the routing of information via a large number of intermediate processors. This broad division is shown in Figure 1.

Switched systems are further subdivided in Figure 2 into those in which all the memory is distributed amongst the processors as local memory (Distributed Memory Switched MIMD), and those in which the main bulk memory is a common shared resource (Shared Memory Switched MIMD). In the former the processor plus local memory forms a processing element (PE), and the PEs communicate through the switch. In the latter all the processors access the shared memory through the switch. A further subdivision is then possible according to the topology of the switch, and examples are given in Figure 2 of Cross-Bar, Multi-Stage and Bus connections in both Shared and Distributed memory systems. Many large systems have both shared common memory and distributed local memory. Such systems may be considered as hybrids or simply as Switched MIMD. Almost all processors have some local memory, however the classification is based on the location of the main bulk memory of the system.

Shared Memory computers appear to be the most common Switched systems, in particular all the current generation of commercial supercomputer fall into this category (Cray X-MP, Cray-2, ETA-10, IBM 3090). In the case of these computers the details of the switch connections between the processors and the memory banks are not generally revealed. However, in the case of more experimental systems sufficient details are available to classify the designs by the topology of the switch. The S1 at Livermore and the C.mmp at CMU used a complete Cross-Bar switch to make possible a direct electrical connection between any of 16 processors and any of 16 memory banks. The MIDAS at the University of California, Berkeley also has a functional crossbar connection. Complete crossbar connections appear to be too expensive at present for more than about 16 processors, and systems that are planned to expand past this number inevitably use multi-stage switches, usually based on variations of the omega switch. Examples are the ULTRA at NYU (a version of which is being implemented by IBM at their Yorktown Heights laboratories as the RP3), and the Cedar at the University of Illinois. On the other hand TRAC at the University of Texas uses a banyan network as a switch. Although the individual processors (PEMs) of the Denelcor HEP are examples of Pipelined MIMD, the overall architecture of 16 PEMs connected through a switch to 128 memory modules (DMMs), falls into the Shared Memory Switched MIMD category. Attaching all the processors and all the memory modules to a common bus is also a popular architecture because of its economy, although the common bus clearly forms a bottleneck preventing such systems expanding to a large size. Examples are the MP/C, Minerva, and the commercially available ELXI 6400 and FPS-5000 series. Distributed Memory Switched systems are much less common but examples are listed in Figure 2 of Cross-Bar, Multi-Stage and Bus systems.

All MIMD Networks appear to be distributed memory systems, but they may be further subdivided according to the topology of the network, as is shown in Figure 3. The simplest Network is the Star, in which several computers are connected to a common host as in the LCAP configuration of IBM at Kingston and Rome. In this loosely coupled array of processors ten FPS-164 computers are connected over channels to an IBM 4381 host. Single and Multi-dimensional meshes are exemplified by the Cyberplus 1D ring design, and the 2D meshes of the NASA Finite Element Machine (FEM) and the University of Columbia VFPP. The ICL DAP, Goodyear MPP and Illiac IV are examples of 2D meshes but they are controlled by a single instruction stream, and are therefore SIMD rather than MIMD computers. Binary Hypercube Networks in which there are only two PEs along each dimension form an interesting class which is receiving a lot of attention in the Cosmic Cube of CIT and its commercial derivatives, the Intel iPSC and the FPS T-machine. Examples also exist of Hierarchical Networks based on trees, pyramids (the University of Erlangen EGPA), and bus connected clusters of processors (Cm*). The most suitable computer network will certainly depend on the nature of the problem to be solved, hence it is attractive to have an MIMD Network which may be reconfigured under program control. The CHiP computer is designed to satisfy this requirement.

3. Vector (SIMD) Performance - the $(r_\infty, n_{1/2})$ benchmark

The first benchmark was developed in order to characterize the performance of vector pipelines [3]. Before the advent of such pipelines, computer performance was measured by a single parameter, namely Mips (millions of instructions executed per second). However, since a single vector instruction initiates differing amounts of work depending on the lengths of the vectors involved, the unit of performance had to be changed to Mflop/s (millions of floating-point operations per second). Further, the fact that vector pipelines have a significant startup time, not present in unpipelined computers, leads to the necessity of introducing a second parameter. One way of doing this uses the two parameters $(r_\infty, n_{1/2})$ as the following analysis shows:

The time, t, for the execution of a vector instruction consists of a startup time, t_0, and a time per element of the vector, t_e multiplied by the number of elements, n.

$$t = t_0 + t_e n \tag{1}$$

The average time per element, $\bar{t}_e$ is then

$$\bar{t}_e = t/n = t_e + t_0/n \tag{2}$$

and the average processing rate, or performance r, is

$$r = \bar{t}_e^{-1} = \frac{1}{(t_e + t_0/n)} \tag{3}$$

After a little rearrangement these formulae can be written

$$t = r_\infty^{-1}(n + n_{1/2}) \tag{4}$$

$$r = \frac{r_\infty}{(1 + n_{1/2}/n)} \tag{5}$$

$$\text{where} \qquad r_\infty = t_e^{-1} \qquad \text{and} \qquad n_{1/2} = t_0/t_e \tag{6}$$

The significance of the parameters r_∞ and $n_{1/2}$ in equation (5) is as follows. In this model of a vector pipeline, the maximum performance occurs for infinitely long vectors and is designated by r_∞. This performance is never quite achieved for vectors of finite length, but is approached asymptotically as the vector length increases. One can tell how close the performance will be to the asymptote by comparing the vector length with the second parameter $n_{1/2}$. If the vector length equals $n_{1/2}$, then the pipeline performance will be half of the asymptotic maximum of r_∞. The performance for other vector lengths can be computed simply from equation (5). The functional form of this approach to the asymptote will occur repeatedly in the subsequent discussion of computer performance, and we define it as the *pipeline* function

$$\text{pipe}(x) = \frac{1}{1 + x^{-1}} = \frac{x}{1 + x} \tag{7}$$

The performance for a vector pipeline can then be written

$$r = r_\infty \text{ pipe}(n/n_{1/2}) \tag{8}$$

It is important to realise that the pipeline function represents a very slow approach to the asymptote. Vectors of length $4 \times n_{1/2}$ are required to achieve 80 % of the maximum performance, of length $9 \times n_{1/2}$ to achieve 90 %, and $99 \times n_{1/2}$ to achieve 99% of the maximum performance.

In this characterization of performance, the parameters r_∞ and $n_{1/2}$ have been chosen rather than the original time parameters t_0 and t_e because they are more closely related to facts about a problem that are known to the computer user. The parameter $n_{1/2}$ provides the user with a yardstick with which to compare the vector lengths that occur in his problem, and r_∞ provides a target with which to compare the actual performance of his program. The characterization also nicely separates the effects of changes in technology used to implement an architecture, from changes to the architecture itself. The former (such as a reduced clock period) only affect r_∞, whilst the latter (such as a change in the number of pipeline segments) only affect the $n_{1/2}$.

The values of $(r_\infty, n_{1/2})$ can be obtained by measuring the time for the following simple loop as a function of the loop length N

$$DO \ 10 \ I = 1, N \tag{9}$$
$$10 \quad A(I) = B(I) * C(I)$$

This loop will be replaced by a vector instruction by any vectorizing compiler. A best straight line is fitted to the measurements of time against vector length. The inverse slope of this line is the value of r_∞, and the negative of the intercept with the N-axis is the value of $n_{1/2}$. Of course, different kernels may be used in the DO-loop, and a table of different performance figures obtained. There is value, however, in adopting the above element-by-element vector multiply kernel as a standard fixed point of comparison between computers, variations about which can be studied separately. As an example, Figure 4 shows measurements obtained on one CPU of the Cray X-MP for three different kernels [7].

4. Parallel (MIMD) Performance - the $(\hat{r}_\infty, f_{1/2}, s_{1/2})$ Benchmark

A simple model of MIMD computing that covers many real programs is based on the FORK/JOIN primitives. A program is considered as a sequence of serial and parallel sections, in which the work is performed respectively by one or many processors (that is to say instruction streams or processes). Each parallel section starts with a FORK which assigns work among the p processors, and initiates the parallel computation in each processor more or less at the same time. Each parallel section ends with a JOIN at which the processors wait until they have all finished their share of the computational work. This pattern of computation can be implemented by FORK/JOIN instructions, or alternative instructions such as the BARRIER. The latter, like JOIN, is a synchronization instruction that makes processors wait until all processors have reached the BARRIER instruction, before any are allowed to proceed. It is important to know the computational cost of the synchronization software and hardware that is provided on any parallel computer system, because this will influence the way in which a particular program is reorganized for parallel execution (i.e. how it is parallelized).

A benchmark has been defined for measuring the synchronization overhead. This involves measuring the execution time t for a single parallel section and fitting it to the expression

$$t = r_\infty^{-1}(s + s_{1/2})$$ (10)

where s is the total amount of computational work in the parallel section measured, for example, in floating-point operations, and $s_{1/2}$ is the overhead of synchronization, measured in terms of the amount of work that could have been done during the time of the synchronization. This is the amount of *lost* work, and measures the cost of synchronization in a currency that has a known value to the programmer. Parallel programming does not come for free, because synchronization is an additional overhead introduced by parallel programming, that is *not* present in serial programming. Therefore $s_{1/2}$ can also be regarded as measuring the cost of going parallel. The parameter r_∞ is the asymptotic performance (measured in Mflop/s if t is in microseconds), which is approached as the amount of work s goes to infinity. The job chosen for the parallel section in the standard benchmark is an element-by-element vector multiply spread as evenly as possible over the p processors. The time for synchronization includes the time to calculate the indices to be used by each processor. Furthermore, in order to vary the amount of arithmetic performed per data item transferred, the arithmetic is repeated an appropriate number of times.

As with the other "subscript 1/2" parameters, $s_{1/2}$ is the amount of work in a parallel section that is necessary to achieve half of the asymptotic performance. Clearly it is hardly worth making a parallel section if the amount of work involved is much less than $s_{1/2}$ floating-point operations, and really one would like to see parallel sections with $s \gg s_{1/2}$ if the parallel hardware is to be efficiently used. The amount of work s in a parallel section is often referred to as the grain of the program parallelism, hence $s_{1/2}$ quantifies the required grain size for the efficient use of a particular parallel system. The performance at other grain sizes can be computed from

$$r = \frac{s}{t} = \frac{r_\infty}{(1 + s_{1/2}/s)} = r_\infty\, pipe\,(s/s_{1/2})$$ (11)

The parameters r_∞ and $s_{1/2}$ are composite numbers depending on many properties of the parallel computer hardware and software. They are intended to show in a simple way and at a high level, approximately how the parallel environment appears to the programmer, using variables that can be related to some properties of the user program. They are not intended to be highly accurate descriptions of the low level timing behaviour of a parallel system, which is likely to be far too complex to describe accurately in such a simple way.

A simple model for the time behaviour of the benchmark is

$$t = t_0(p) + t_i(p)\,m + t_a(p)\,s$$ (12)

$$t = t(startup\ and\ synchronization) + t(communication) + t(calculation)$$ (13)

where the three terms are identified respectively with startup and synchronization, communication and calculation as shown, and

s total number of floating-point operations (flop) performed in all processors

m the number of I/O data words in the parallel section (see below)
$f = s/m$ floating point operations per I/O data word
$t_0(p)$ time for the null job ($s = m = 0$)
$t_t(p)$ time per I/O word on average
$t_a(p)$ time per floating point operation on average

The problem variable m quantifies the communication that a parallel section has with the rest of the program. If, as is usual, the body of a parallel section is written as a subroutine, it is the number of words contained in all the input and output variables and arrays of the subroutine, for all instantiations to the subroutine. In the case of the above benchmark, if the vector length is n, then $m = 3n$ (two input vectors and one output vector), and $s = 3nf$, because the arithmetic is repeated $3f$ times.

Equation (12) can be rewritten in form of equation (10), and we obtain the benchmark parameters in terms of the time parameters

$$r_\infty = \frac{\hat{r}_\infty}{(1 + f_{1/2}/f)} = \frac{1}{(t_a + t_t/f)} \tag{14}$$

$$s_{1/2} = t_0 r_\infty = \frac{\hat{s}_{1/2}}{(1 + f_{1/2}/f)} = \frac{t_0}{(t_a + t_t/f)} \tag{15}$$

where $\quad \hat{r}_\infty = t_a^{-1}, \quad \hat{s}_{1/2} = t_0/t_a, \quad$ and $\quad f_{1/2} = t_t/t_a \tag{16}$

Using these variables the timing equation (11) can be rewritten

$$t = t_0(p) + \frac{f_{1/2}}{\hat{r}_\infty}m + \frac{s}{\hat{r}_\infty} \tag{17}$$

Thus the ratio of f (a program variable) to $f_{1/2}$ (a hardware variable) determines the fraction of the peak performance $\hat{r}_\infty$ that is obtained. The peak performance being that obtained in the absence of any communication delays ($f_{1/2} = 0$), and reflects the bare performance of the arithmetic hardware in a single processor. In the present context of serial versus parallel processing, we would observe an asymptotic performance equal to the peak performance when executing the program in serial mode in the processor that contains the data array. However, in parallel mode we would find a reduced asymptotic performance (perhaps much reduced) given by (14) depending on the relative values of f and $f_{1/2}$, because the data must be communicated to the parallel processors. In the standard vector multiply benchmark program, there are three data transfers for every multiply operation (two input arguments and one result), hence $f = 1/3$. Therefore the vector multiply operation is repeated $3f$ times in the benchmark.

Combining equations (11) and (14) we obtain the following expression for the overall performance

$$r = \frac{\hat{r}_\infty}{\left(1 + \dfrac{s_{1/2}}{s}\right)\left(1 + \dfrac{f_{1/2}}{f}\right)} = \hat{r}_\infty \, pipe\left(\frac{s}{s_{1/2}}\right) pipe\left(\frac{f}{f_{1/2}}\right) \tag{18a}$$

$$= \frac{\hat{r}_\infty}{\left\{ 1 + \frac{f_{1/2}}{f} + \frac{\hat{s}_{1/2}}{s} \right\}} \tag{18b}$$

where the first factor in the denominator (or pipe function) of equation (18a) gives the degradation of performance due to startup and synchronization, and the second factor the degradation due to communication delays.

As an example of the benchmark we include the results obtained on the LCAP system at the IBM European Center for Scientific and Engineering Computing (ECSEC, Rome) [5]. The system comprises an IBM 4381 host attached by channels to ten FPS-164 computers, and the benchmark is realised by holding the data arrays in the host and transferring approximately $1/p$th of the vectors to each of p processors. After computation in the FPS-164s, the partial result vectors are returned to the host and assembled. The host controls the calculation but for simplicity of interpretation does not perform any of the arithmetic (although of course it could). Synchronization takes place on the host using the VM/EPEX parallel programming system and the BARRIER instruction [6]. Timing is done on the host using the system time-of-day clock to measure the elapsed "wall clock" time for the job.

The startup and synchronization time is found to vary approximately linearly with the number of processors

$$t_0(p) = 2500 + 4777 p \qquad \mu s \tag{19}$$

Figure 5 shows the variation of the asymptotic performance r_∞ as a function of p for the extreme cases of $f = 1/3$ and $f = 100$. The dotted lines show the fit obtained using equation (14) with the parameters

$$\hat{r}_\infty = 1.08 \, p \text{ Mflop/s}, \quad \text{and} \quad f_{1/2} = 2.02 \text{ flop/word} \tag{20}$$

Similarly Figure 6 shows the variation of $s_{1/2}$ using the same parameters and equation (15). The above parameters give rise to the complete timing formula

$$t = 2500 + 4777 p + \frac{1.87}{p} m + \frac{0.926}{p} s \qquad \mu s \tag{21}$$

thus

$$t_t(p) = \frac{1.87}{p} \, \mu s/\text{word}, \quad \text{and} \quad t_a(p) = \frac{0.926}{p} \, \mu s/\text{flop} \tag{22}$$

The fact that we observe the communication term $t_t(p)$ inversely proportional to the number of processors p, shows that the channels are working in parallel as expected. Since the data arrays are transferred as 4-Byte numbers across the channels, the observed data rate corresponds to 2.14 MByte/s per FPS processor on average. This is consistent with the hardware configuration that provides between 1 and 3 MByte/s per processor depending on what combination of FPS-164s are used.

The above values of $f_{1/2}$ and $s_{1/2}$ for the LCAP show clearly what are the characteristics of a problem that will be suitable for running on this parallel computer system. With an $f_{1/2} = 2$ flop it is clear that the problem must be such as to have at least 10 floating-point operations per I/O operation to the host. This would give an asymptotic performance slightly better than 80 % of the peak performance. However the fraction of the asymptotic performance realised depends on the grain size s compared to $s_{1/2}$ through equation (11). Figure 6 shows that grain sizes between a hundred thousand and a million floating-point operations would be necessary to achieve a reasonable percentage of the asymptotic performance. This quantifies the meaning of a satisfactory grain size for successful parallel programming on the LCAP, and is the reason that the system is described as a loosely coupled array of processors. For comparison, values $s_{1/2}$ obtained on a single PEM of the Denelcor HEP using up to 20 instruction streams ranged from 200 to 800 flop depending on the software used [4].

5. Conclusions

In this paper we have given a classification of vector and parallel computer systems, and shown how their performance can be characterized by the use of a small number of parameters. These parameters give some answers to such questions as: "What is a long and short vector?", "What grain size is necessary for efficient parallel computation?", "What are the costs of synchronization and communication?", "How tightly coupled is this parallel system?".

6. References

[1]   M. J. Flynn
        "Some computer organizations and their effectiveness"
        IEEE Trans. Comput., C-21, (1972), pp.948-960

[2]   R. W. Hockney
        "MIMD computing in the USA - 1984"
        Parallel Computing 2 (1985), pp.119-136

[3]   R. W. Hockney and C. R. Jesshope
        Parallel Computers - Architecture, Programming and Algorithms
        Adam Hilger (Bristol 1981)

[4]   R. W. Hockney "Performance characterization of the HEP" in
        MIMD Computation: HEP Supercomputer, (J. S. Kowalik,ed.), MIT Press, (Cambridge, Mass. 1985), pp.59-90

[5]   R. W. Hockney
        "Synchronization and communication overheads on the LCAP multiple FPS-164 computer system"
        Parallel Computing, to appear

[6]   F. Darema-Rogers, D. A. George, V. A. Norton and G. F. Pfister
        "VM/EPEX - A VM environment for parallel execution"
        IBM Research Report RC11225 (#49161), Yorktown Heights, NY, 1/23/85

[7]   R. W. Hockney
        "$(r_\infty , n_{1/2} , s_{1/2})$ measurements on the 2-CPU CRAY X-MP"
        Parallel Computing 2 (1985), pp.1-14

7. Figures

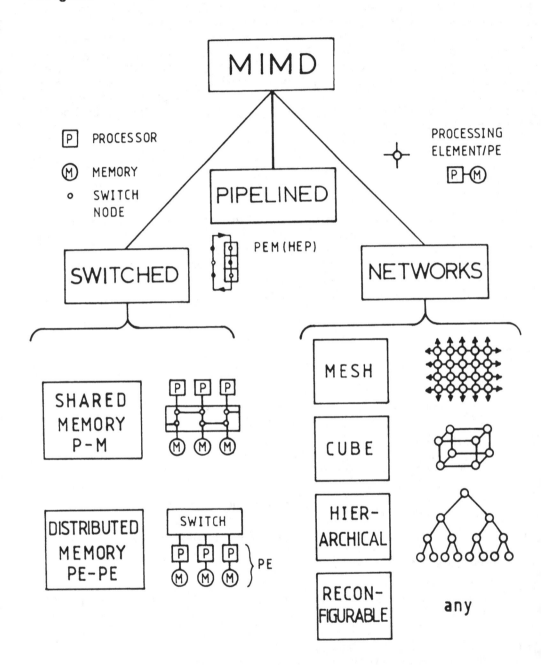

Figure 1. The overall classification of MIMD computer systems into Pipelined, Switched and Network systems (*from Hockney* [2], *courtesy of North-Holland*)

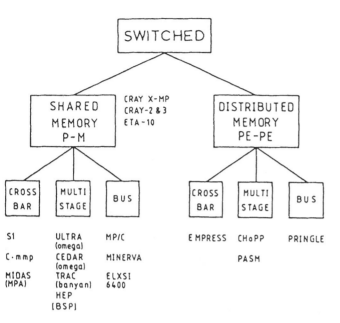

Figure 2.   Subdivision of Switched MIMD computers according to the allocation of memory and the type of switch used (*from Hockney* [2], *courtesy of North-Holland*)

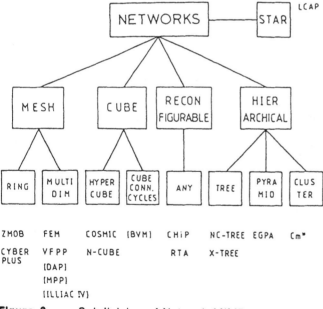

Figure 3.   Subdivision of Network MIMD computers according to the topology of the network (*from Hockney* [2], *courtesy of North-Holland*)

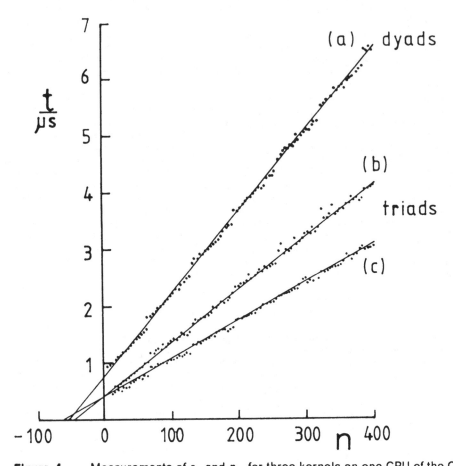

Figure 4. Measurements of r_∞ and $n_{1/2}$ for three kernels on one CPU of the Cray X-MP.

The asymptotic performance, r_∞, is the inverse slope of the best straight line fit to the time t versus vector length n data. The half performance length, $n_{1/2}$, is the negative intercept of the line with the n axis.

(a) $A = B \cdot C$ all vector dyadic operation: $r_\infty = 70$ Mflop/s, $n_{1/2} = 53$.
(b) $A = D \cdot B + C$ all vector triadic operation: $r_\infty = 107$ Mflop/s, $n_{1/2} = 45$.
(c) $A = s \cdot B + C$ where s is a scalar (Cyber 205 triad): $r_\infty = 148$ Mflop/s, $n_{1/2} = 60$.

(from Hockney [7], courtesy of North-Holland)

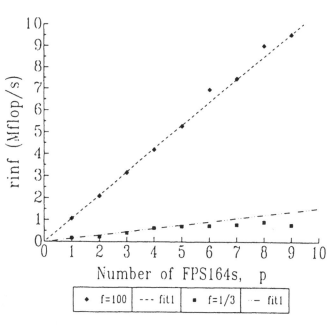

Figure 5. Measurements of asymptotic performance r_∞ (here written rinf) as a function of the number of FPS-164 processors on the LCAP computer, for two values of f. Fit 1 is the theoretical prediction using equation (14) and the parameters (20).

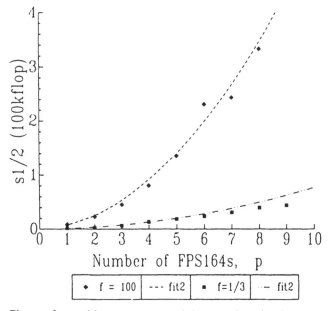

Figure 6. Measurements of the synchronization overhead parameter $s_{1/2}$ as a function of the number of FPS-164 processors on the LCAP computer, for two values of f. Fit 2 is the theoretical prediction using equation (15) and the parameters (19) and (20).

Part II - Parallel Computer Architectures

Shared Memory, Vectors, Message Passing, and Scalability

Burton J. Smith

Supercomputing Research Center
4380 Forbes Blvd.
Lanham, MD 20706, USA

Abstract

The many zealots of parallel computation are split into several factions, each believing that it has the best (or sometimes, only) possible architectural idea for the fast computers of the future. Comparisons between the ideas have been rare, possibly because the zealots don't like to acknowledge the merits of each other's architectures. This paper compares notions from two popular types of MIMD [Flyn72] parallel computers: the message passing systems with many processors like those manufactured by Ncube and Intel, and the traditional shared memory systems of a few vector processors, Cray's products for example. Messages are compared with vectors, sending and receiving a message is compared with loading and storing a vector, and the idea of using parallelism to keep a waiting processor busy is compared with the idea of using parallelism to keep a vector pipeline busy. The conclusions are that a good shared memory system and a good message passing system are indistinguishable, and that parallelism is used in both kinds of systems not just to translate increased hardware into increased speed but also to achieve good hardware utilization in spite of nonlocality in space or time.

1. Introduction

The many zealots of parallel computation are split into several factions, each believing that it has the best (or sometimes, only) possible architectural idea for the fast computers of the future. Comparisons between the ideas have been rare, possibly because the zealots don't like to acknowledge the merits of each other's architectures. This paper compares notions from two popular types of MIMD [Flyn72] parallel computers: the message passing systems with many processors like those manufactured by Ncube and Intel, and the traditional shared memory systems of a few vector processors, Cray's products for example.

The nonshared memory advocates usually claim that typical parallel programs have abundant parallelism, and the key challenge is to map, i.e. *schedule*, both the data structures and computations onto the computer. Their viewpoint is wholly understandable. Nonshared memory systems incorporating thousands or tens of thousands of processors are feasible. When compared to shared memory architectures, however, their kind of architecture is substantially more difficult to schedule computations for, either in space, by mapping data structures onto memories, or in time, by mapping computations onto processors [LuFa87].

On the other hand, the vector and shared memory advocates believe that the typical parallel program has only a modest amount of parallelism, and that the real challenge of parallel programming is discovering or creating this parallelism. Their viewpoint derives from their experiences with parallel program design for current shared memory systems, and the relative simplicity of the scheduling problem once enough parallelism of an appropriate kind has been found.

There is merit in the points of view of both factions. Significant improvements in the breadth of application of general purpose high speed parallel computing will require easy-to-schedule systems with many processors and much better techniques for writing parallel programs. This paper compares some of the key attributes of the two kinds of system, with the idea that a synthesis may be possible in an architecture that both passes messages and shares memory.

2. Messages and Vectors

What are the differences between messages and vectors? In this context, "message" means a sequence of values passed between two processors. The values in the sequence may be of different types, a mixture of integers and floating point numbers for example. When a message is sent from one processor to another in a nonshared memory system, the values in the message are copied one after another from the memory of the sending processor to the memory of the receiving processor. High performance nonshared memory systems based on message passing use special hardware that can transfer entire messages without software intervention. In these systems, messages are usually stored in consecutive memory locations to simplify the hardware design. In shared memory systems, neither copying nor consecutive storage is necessary, but it may be desirable to copy a message anyway to avoid future memory conflicts and thereby obtain more parallelism.

Synchronization between a message transfer and a processor is typically done with a *blocking receive*; when a processor wants to receive a message, it calls a function that returns when the message has arrived. To avoid wasting processor cycles, it is important that the receiving processor have some other work to do while it is waiting. These alternative computations can be generated by decomposing a problem into more parallel parts than would ordinarily suffice for the number of physical processors available. Each parallel part is then assigned to a different *virtual processor* (in this context often called a *task* or *process*). When a virtual processor waits for a message, the physical processor just switches to another virtual processor and keeps on computing.

The time needed to switch between virtual processors is important to the programmer because it determines the maximum message rate that the system will support. If a decomposition of a problem into parallel parts results in too few instructions executed per message sent, then the performance of each physical processor in the system will be limited by the peak message rate. In such cases, the system overhead of a virtual processor implementation is too great for the proposed problem decomposition, and another decomposition should be sought that requires less virtual processor switching. In other words, systems with *lightweight* (easily switched) virtual processors are more generally applicable than are systems with *heavyweight* (slow to switch) virtual processors.

At first glance, a vector looks a lot like a message. A vector is also sequence of values, but unlike a message that is passed between two processors, a vector is passed between a processor and a memory. In all vector systems built so far, each processor manipulates vectors within a single address space. In other words, a vector in a nonshared memory system must be wholly located in the local memory of a single processor, just as a message is. It is possible to imagine a nonshared memory vector architecture in which a processor can load or store a vector in either its own or any other processor's memory with equal ease, but no such system has been built so far. In a similar vein, although it seems to be a distinction that vectors need not be stored in strictly consecutive memory locations as messages are, there is nothing to prevent a more flexible implementation of messages that would remove the distinction by allowing a message to be stored at other than unit stride or even "gathered" at its source and "scattered" at its destination.

There is an important distinction between vectors and messages, however. The values making up a vector are constrained to be of the same type so that vector pipelining is possible. In fact, a vector is really more like a sequence of messages rather than a single one, and each element of the vector is a constituent of an independent parallel computation, performed by a distinct virtual processor. It is this idea that makes delay due to memory and function unit latency relatively benign for vectors in a pipelined system. It is also this very idea that causes arguments as to whether a single vector pipelined processor should be thought of as having SISD or SIMD architecture. A virtual processor here is extremely simple; its only task is to do one of the scalar operations that together make up the vector operation. These are lightweight virtual processors indeed, and the motivation is the same as before, namely finding some other work for a physical processor to do while it is waiting.

The virtual processor idea is at least as old as multiprogramming, but it is especially relevant to parallel computing because it is only the presence of parallelism that makes virtual processors possible. One of the great strengths of the idea is that the exact reason a physical processor is waiting need not matter. If the virtual processors are lightweight enough, a single mechanism can handle waiting due to function unit latency, memory latency, interconnection network latency, memory interference, or message waiting. The important properties of a virtual processor implementation are its flexibility in handling these different causes of delay in parallel computations and the interaction of the virtual processor implementation with the architecture and the programming of the rest of the system.

3. Shared Memory

Why are shared memory MIMD computers with many processors difficult to implement? The answer depends in part on the definition of the term "shared memory", but there is probably broad agreement that any sort of shared memory system with hundreds of processors is a challenge. The main problem is the latency associated with memory access and its consequences for processor performance. If many processors are sharing memory, propagation delays in the interconnection network through either logic circuitry or wiring will limit the minimum latency attainable. There are two possible solutions to this problem, namely latency avoidance and latency tolerance. Latency avoidance is accomplished by arranging a processor's memory accesses so that most of them are to locations that are both spatially and temporally nearby. Latency tolerance has already been mentioned in connection with vectors and messages; it is brought about by introducing virtual processors and additional parallelism. Both ideas have been used in shared memory systems, with varying success.

The most popular method for latency avoidance places a cache between each processor and the interconnection network [GKLS83] [PBGH85]. A cache *line* containing several consecutive words is copied from the shared memory to the cache whenever a miss occurs, i.e. whenever a copy of the addressed word is not presently in the cache. Although a cache miss provides no reduction in latency and consumes more memory bandwidth than an ordinary memory access would, the hope is that the cached word, and other nearby words fetched as part of the cache line, are likely to be referenced in subsequent cycles. If the resulting probability of a cache miss is sufficiently low, the average latency will be decreased; if a word, once cached, is referenced several times, then the shared memory bandwidth needed will be reduced as well.

There are two primary difficulties with caches in shared memory MIMD systems [LeYL87]. First, a cache line can be viewed as a vector whose components are necessarily stored in consecutive memory locations. Random addressing patterns or even moderate strides are penalized by frequent cache misses, perhaps as often as once per memory reference. The result is that the cache line size should be small, perhaps only a single word, and preloading the cache is essentially mandatory. A more serious problem with caches can be inferred by noticing that the caches contain copies of the data rather than the data itself. The processors are reading and sometimes modifying these copies, and yet the data in shared memory should appear consistent. The common definition of consistency is the "as if serial" one: the processors should behave as if they were sequentially reading and writing the original data, a word or perhaps a line at a time, rather than reading and writing several copies of it simultaneously. Trying to maintain cache *coherence* by deciding after each memory write which of the caches now have invalid lines and then arranging that all those caches get the new version is almost certainly impractical in systems with many processors. A cache designed and used in a way that circumvents both difficulties, by prefetching one-word cache lines and by scheduling cache contents to avoid any shared writable data, is essentially a local memory. In other words, the shared memory has become nonshared to some extent.

Latency tolerance is an alternative to latency avoidance, and is a natural benefit of virtual processing by vector pipelining. Unfortunately, vector gathers and scatters are not completely compatible with shared memory MIMD computation because unpredictable memory interference disrupts the smooth flow of vectors between memory and processor. If a single processor is loading a vector from memory, it is relatively easy for the hardware to predict the arrival rate and to guarantee that the order of the vector components is the same as the order in which its addresses were generated. In contrast, when a multitude of processors simultaneously gathers or scatters values at irregular and conflicting memory locations, both the rate and the order in which values arrive at a destination will be irregular as well. This difficulty probably rules out "memory-to-memory" vector operations in shared memory architectures, and may make chaining less effective or even infeasible for vector register based systems. The absence of chaining in a vector processor implies that more parallelism (and more vector registers) will be needed to keep such a processor busy.

The context switching used to obtain good processor utilization while waiting for messages in a message passing computer is not normally thought of as a way of tolerating memory latency. In dataflow architectures [ArCu86] the context switching is so frequent, namely after every operation, that it does precisely that. The original motivation was a desire to exploit fine-grain parallelism, especially parallelism within scalar arithmetic expressions. This type of parallelism generates frequent message delays of short duration, and the dataflow solution to the "keeping busy" problem is to introduce virtual processors. The serendipitous additional benefit, the ability to tolerate memory latency, has turned out to be at least as important a consequence. The HEP computer [Smit78] was similar to dataflow computers in several respects, including this one. The use of these extremely lightweight virtual processors makes it possible to both share memory *and* pass messages efficiently in a single architecture. The vector derangement difficulties suffered by vector processors are not a problem here; as a matter of fact, the message delays that were the motivation for this particular sort of virtual processor machinery are inherently highly irregular. Irregular memory latency and irregular message delay are dealt with identically in these designs.

4. Scalability

What does it mean for a parallel architecture to be scalable? It is certainly too strict to require that arbitrarily large systems be constructible from sufficient quantities of fixed processor, memory, and switch components. It is probably even too much to insist on a fixed instruction set since this would imply indefinite length addresses to accommodate indefinitely large processor or memory configurations. The general idea, though, is that system size should be expandable by a few orders of magnitude without changing the components, and a few orders more without changing the architecture.

A design must be more than just physically expandable to be scalable; it should also be suitable for the same kinds of computations in its large configurations that it is in its small ones. It is therefore not enough to require only that *some* computation or other obtains high speed on the largest of the configurations. The honest criterion is that an unmodified program, if given a bigger computation and executed on a bigger system, should exhibit a respectable speed improvement. It is obviously important that a parallel computing system be scalable in this sense to permit evolutionary rather than revolutionary development of application software, system software, and hardware.

The interesting question to ask about such a scalable system (whether real or hypothetical) is not the one usually asked, namely, "Given identical increases in both system size and problem size, how does speed increase"? It is often wrongly argued that any answer other than "linearly" is a confession of nonscalability. The argument is wrong because the question presumes too much about the relationship among these factors; to get linear speedup, problem size should grow somewhat faster than system size. This extra growth is needed to provide additional parallelism for the extra virtual processors, processors that are needed in turn to overcome the effects of increasing memory latency and message delay. Therefore, the question should be, "Given identical increases in both system size and performance, how must problem size increase"? In a vector pipelined architecture, the vectors will have to be longer; in a message passing system, increased trans-

mission times will require a greater workload per processor to keep the processors busy.

5. Conclusions

A good message passing architecture should be able to send and receive short
messages, even single words, frequently and with high efficiency; a good shared
memory architecture should be able to avoid the need for rigorous scheduling by
being able to freely exchange parallelism for locality. These objectives can be
attained in a single architecture. Scalable shared memory MIMD computers have
been designed that use extremely lightweight virtual processors to overcome the
effects of function unit latency, memory latency, interconnection network latency,
memory interference, and message waiting all at once. Architectures based on
these ideas greatly simplify the task of both the programmer and the compiler
writer, and are excellent candidates for embodiment in future high speed parallel
computers.

6. References

[Flyn72]   M. J. Flynn
             "Some Computer Organizations and their Effectiveness", *IEEE Transactions
             on Computers* **21**, pp. 948-960 (September 1972).

[LuFa87]   O. M. Lubeck and V. Faber
             "Modeling the Performance of Hypercubes: a Case Study Using the Parti-
             cle-in-cell Application", Los Alamos Computer Research and Applications
             Group Report LA-UR87-1522 (1987).

[GKLS83]   D. Gajski, D. Kuck, D. Lawrie, and A. Sameh
             "CEDAR - A Large Scale Multiprocessor", *Proc. 1983 International Confer-
             ence on Parallel Processing*, pp. 524-529.

[PBGH85]   G. Pfister, W. Brantley, D. George, S. Harvey, W. Kleinfelder, K. McAuliffe,
             E. Melton, V. Norton, and J. Weiss
             "The IBM Research Parallel Processor Prototype (RP3): Introduction and
             Architecture", *Proc. 1985 International Conference on Parallel Processing*,
             pp. 764-771.

[LeYL87]   R. L. Lee, P-C. Yew, and D. H. Lawrie "Multiprocessor Cache Design Con-
             siderations", *Proc. 14th Annual International Symposium on Computer
             Architecture* pp. 253-262 (1987).

[ArCu86]   Arvind and D. E. Culler "Dataflow Architectures", *Annual Review of Com-
             puter Science 1986* **1**, pp. 225-253.

[Smit78]   B. J. Smith "A Pipelined, Shared Resource MIMD Computer", *Proc. 1978
             International Conference on Parallel Processing*, pp. 6-8.

Communication Techniques in Parallel Processing

Howard Jay Siegel
Supercomputing Research Center
4380 Forbes Blvd.
Lanham, MD 20706, USA

William Tsun-yuk Hsu
Center for Supercomputing Research & Development
University of Illinois
Urbana, IL 61801, USA

Menkae Jeng, Wayne G. Nation
PASM Parallel Processing Laboratory
School of Electrical Engineering
Purdue University
West Lafayette, IN 47907, USA

Abstract

Large-scale parallel processing is one basis for the design of the supercomputer systems needed for many scientific, industrial, and military applications. The interconnection network in a parallel processing system provides the vehicle for communications among the processors and memories. Eight interconnection techniques for supporting large-scale parallelism (e.g., 2^6 to 2^{16} processors) are overviewed. These are the Cube, Shuffle-Exchange, PM2I (Plus-Minus 2^i), and FNN (Four Nearest Neighbor) single stage networks, and the Generalized Cube, Extra Stage Cube, Augmented Data Manipulator, and Dynamic Redundancy multistage networks.

1. Introduction

A large amount of processing power is required by many of today's scientific, industrial, and military problems; e.g., aerodynamic simulations, air traffic control, ballistic missile defense, biomedical image processing, chemical reaction simulations, fluid dynamics studies, map making, seismic data processing, satellite-collected imagery analysis, missile guidance, plasma physics simulations, robot vision, speech understanding, and weather forecasting. To do these applications there is a computational "need for speed" due to time constraints on attaining the solution combined with the computational complexity of the algorithms involved and/or the sizes of the data sets to be processed.

This project was initiated while the authors were at Purdue University. Some of the material in this paper is summarized from H.J. Siegel, *Interconnection Networks for Large-Scale Parallel Processing*, Lexington Books, D.C. Heath and Company, Lexington, Mass., USA, copyright 1985.

Parallel processing systems comprising a multitude of tightly-coupled processors, working together to execute a single overall task, can help provide the performance required for these computations. This paper overviews some techniques for providing the communications among the processors and memories of large-scale parallel systems.

The *processor-to-memory* model of a parallel computer system (Figure 1) assumes N processors on one side of a bidirectional network and N memory modules on the other side. In the *PE-to-PE* model (Figure 2), PE i is connected to input i and output i of a unidirectional interconnection network, where each *PE* (processing element) is a processor paired with its own memory. The PE-to-PE model will be used in this paper; however, the material presented is also applicable to processor-to-memory systems.

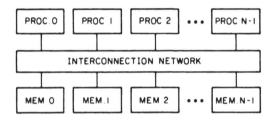

Figure 1. Processor-to-memory parallel machine configuration with N processors and N memories.

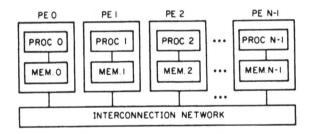

Figure 2. PE-to-PE parallel machine configuration with N processing elements (PEs).

There are different modes of parallelism [Fly66].

An **SIMD** (single instruction stream - multiple data stream) machine consists of N PEs, an interconnection network which provides communications between PEs, and a single control unit (Figure 3). The control unit broadcasts instructions to all the PEs, and all enabled PEs execute the same instructions simultaneously. Each PE operates on its own data from its memory. Examples of SIMD machines that have been built are the Illiac IV [BoD72], STARAN [Bat77], MPP [Bat82], and the Connection Machine [Hil85]. IBM's GF11 [BeD85] is an SIMD machine currently under construction.

An **MIMD** (multiple instruction stream - multiple data stream) machine consists of N PEs linked by an interconnection network (Figure 2). Each PE stores and executes its own instructions and operates on its own data. Examples of

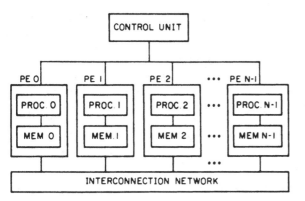

Figure 3. SIMD machine configuration with N processing elements (PEs).

MIMD machines that have been built are Cm* [SwF77], the BBN Butterfly [CrG85], the Cosmic Cube [Sei85], and HEP [HaD85]. MIMD systems under development include the NYU Ultracomputer [GoG83] and the IBM RP3 [PfB85].

Multiple-SIMD machines include a set of control units and can be reconfigured into a number of smaller, independent SIMD machines of various sizes. Proposed multiple-SIMD systems are MAP [Nut77] and the original design for the Illiac IV [BaB68].

Partitionable SIMD/MIMD machines can be partitioned into smaller independent machines of different sizes working in SIMD and/or MIMD mode. Examples of partitionable SIMD/MIMD systems in the prototype stage are PASM [SiS81, SiS87] and TRAC [SeU80].

The interconnection scheme for an N-PE parallel system, where N is 2^6 to 2^{16}, must provide fast and flexible communications without unreasonable cost. A single shared bus is not sufficient, since it is often desirable to allow all PEs to send data to other PEs simultaneously (e.g., for PE i to PE i+1, $0 \leq i \leq N - 2$). On the other extreme, to link directly each PE to every other PE is highly impractical for large N, since it requires $N \times (N - 1)$ unidirectional lines. The crossbar network, another approach that allows every PE to communicate with any unique PE simultaneously, can be viewed as N vertical lines intersecting N horizontal lines, where each line intersection is a crosspoint switch [Thu79]. The number of crosspoint switches grows with N^2, which, given current technology, makes crossbars infeasible for large systems. Networks between these extremes have been proposed. There is no one network that is generally considered "best," since the cost-effectiveness of a particular network design depends on such factors as the computational tasks for which the network will be used, the desired speed of interprocessor data transfers, the actual hardware implementation of the network, the number of processors in the system, speed of the processors and memories being interconnected, and any cost constraints on the construction. A variety of networks which have been proposed are overviewed in numerous survey articles and books, e.g. [Bae80, BrH83, Fen81, HoJ81, HwB84, Kuc78, MaG79, Sie85, SiM79, ThM79, WuF84].

This paper is a study of an important collection of network designs that can be used to support large-scale parallelism, i.e., these networks can provide the communications needed in a parallel processing system consisting of a large number of processors (e.g., 2^6 to 2^{16}) which are working together to perform a single overall task. Many of these networks can be used in dynamically reconfigurable machines which can perform multiple independent tasks, where each task is processed using parallelism. Both single stage and multistage networks are examined. In a single stage network, data items may have to be passed through the switches several times before reaching their final destinations. In a multistage network, generally one pass through the multiple (usually $\log_2 N$) stages of switches is sufficient to transfer the data items to their final destinations.

The networks overviewed here are based on the Shuffle-Exchange, Cube, PM2I (Plus-Minus 2^i), and FNN (Four Nearest Neighbor) interconnection patterns. These networks and their single stage implementations are explored in Section 2. Section 3 is a study of the multistage Cube/Shuffle-Exchange class of networks. The Generalized Cube network will be discussed as an example of this type of network. A fault tolerant version of the Generalized Cube network, called the Extra Stage Cube network, is the subject of Section 4. Data manipulator type networks, which are multistage implementations of the PM2I connection patterns, will be described in Section 5. In Section 6, the Dynamic Redundancy network, a data manipulator like network capable of acting as a fault tolerant Generalized Cube network and allowing spare PEs, is examined.

2. Single Stage Interconnection Networks

Assume there are $N = 2^m$ PEs, numbered (addressed) from 0 to $N - 1$. An *interconnection network* can be described by a set of interconnection functions, where each *interconnection function* is a bijection (permutation) on the set of PE addresses [Sie77]. When an interconnection function f is executed, PE i sends data to PE f(i). If a system is operating in SIMD mode and all PEs are active, this means that every PE sends data to exactly one PE, and every PE receives data from exactly one PE. Otherwise, the data transfer may occur only for a subset of the PEs in the system. The *partitionability* of an interconnection network is the ability to divide the network into independent subnetworks of different sizes [Sie80]. Each subnetwork of size $N' < N$ must have all of the interconnection capabilities of a complete network of that same type built to be of size N'. Partitionable networks are needed to support multiple-SIMD and partitionable SIMD/MIMD systems, and can also be used to partition MIMD machines into smaller MIMD virtual machines.

The **Cube** network (Figure 4) consists of m interconnection functions:

$$cube_i(p_{m-1} \cdots p_{i+1}p_ip_{i-1} \cdots p_0) = p_{m-1} \cdots p_{i+1}\bar{p}_ip_{i-1} \cdots p_0$$

for $0 \le i < m$, where $p_{m-1} \cdots p_1p_0$ is the binary representation of an arbitrary PE address P and $\bar{p}_i$ is the complement of p_i [Sie77]. For example, $cube_2(3) = 7$. This network is called the Cube because when the PE addresses are considered as the vertices of an m-dimensional cube, using an appropriate labeling, this network connects each PE to its m neighbors (Figure 5). System designs using the single stage Cube include the Cosmic Cube MIMD machine [Sei85], the proposed

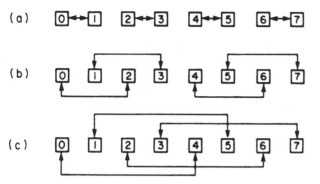

Figure 4.   Cube network for N = 8.
(a) cube₀ connections; (b) cube₁ connections; (c) cube₂ connections.

CHoPP MIMD machine [SuB77], the Intel iPSC MIMD machine [Int85]. and the Connection Machine SIMD system [Hil85]. The Cube network forms the underlying structure of the multistage networks in Sections 3 and 4.

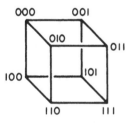

Figure 5.   Cube structure for N = 8, with vertices labeled from 0 to 7 in binary.

When operating in SIMD mode the network settings for data transfers are determined by the control unit. The control unit simply specifies the Cube function to be performed and the set of PEs which would be involved in the transfer, all as part of the SIMD program. A given data movement among the PEs may require a sequence of Cube functions to be executed. For example, for $N = 8$, if all PEs do $cube_1$, and then just PEs 0, 1, 4, and 5 do $cube_2$, the resulting transfer is equivalent to sending data from PE j to PE j+2 mod 8 [Sie79]. In MIMD mode, inter-PE data transfers are less structured. PEs send data through the network independently. There may be several ways to route a message through intermediate PEs. For example, for $N = 8$, to transfer data from PE 4 to PE 2, one possible sequence of $cube_i$ transfers would be $cube_1$ and $cube_2$, moving the data from PE 4 to 6, and then from PE 6 to 2. The order of performing the $cube_i$ functions is not important. For example, $cube_2$ and then $cube_1$ would move data from PE 4 to 0, and then from PE 0 to 2.

The partitioning of the Cube network can be done based on any bit position of the PE addresses. If the use of the *cube*, function is disallowed, all PEs with a '0' in the i-th bit of their addresses cannot communicate with PEs with a '1' in the i-th bit of their addresses. These two groups of PEs can form two independent virtual machines. Two independent subnetworks of size N/2 are formed, each with $m - 1 = \log_2(N/2)$ Cube functions. Each of these Cube subnetworks can then be further subdivided into smaller partitions. In general, the physical addresses of all the PEs in a partition of size 2^s must agree in the $m - s$ bit positions corresponding to the $m - s$ Cube functions the partition cannot use for communications.

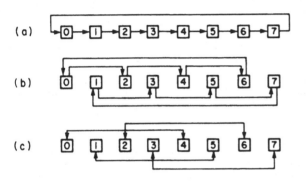

Figure 6.   PM2I network for N = 8.
(a) $PM2_{+0}$ connections; (b) $PM2_{+1}$ connections; (c) $PM2_{+2}$ connections; $PM2_{-i}$ is the same as $PM2_{+i}$ except the direction is reversed.

The **Plus-Minus 2^i (PM2I)** network (Figure 6) consists of 2m interconnection functions:

$$PM2_{+i}(P) = P + 2^i \bmod N$$

and

$$PM2_{-i}(P) = P - 2^i \bmod N$$

for $0 \leq i < m$ [Sie77]. For example, $PM2_{+2}(4) = 8$ if $N > 8$. This network is called the Plus-Minus 2^i since it allows PE P to send data to any one of $PE\ P + 2^i$ or $PE\ P - 2^i$, arithmetic mod N, $0 \leq i < m$ (where $-j \bmod N = N - j \bmod N$). Networks similar to the PM2I are used in the design of the "Novel Multiprocessor Array" [OkT82], the Omen computer [Hig72], and the SIMDA machine [Wes72]. The data manipulator multistage networks in Section 5 are based on the PM2I.

In SIMD mode, network control is as in the Cube network. For example, for $N = 8$, if all PEs do $PM2_{+0}$ and then just the even numbered PEs do $PM2_{-1}$, the resulting transfer is equivalent to doing a $cube_0$ [Sie79]. Data transfers in MIMD mode can be implemented based on the difference between the source and destination addresses. For example, for $N = 16$, to route data from PE 1 to PE 15, 14 has to be added to 1. One sequence of PM2I functions that would perform this transfer is $PM2_{+1}$, $PM2_{+2}$, and $PM2_{+3}$, and the message would go through PEs 1, 3, 7, and 15. As in the Cube network, the order of performing the PM2I functions is not important. Also, a different set of PM2I functions may result in the same routing if the sum of the functions is equal to the difference between the source and destination addresses; e.g., it is also possible to use $PM2_{-1}$ to route data from PE 1 to PE 15.

To partition the PM2I network into two PM2I subnetworks of size N/2, use of the $PM2_{\pm 0}$ interconnection function must be disallowed. This subdivides the PEs into an even-numbered group and an odd-numbered group, and PEs in one group cannot communicate with PEs in the other group. This is because the only way for a PE in the odd-numbered group to communicate with a PE in the even-numbered group is to use $PM2_{\pm 0}$. Each of the two subnetworks formed has $2(m - 1) = 2 \log_2(N/2)$ PM2I functions. Each of these subnetworks can be further divided into smaller subnetworks by disallowing $PM2_{\pm 1}$, $PM2_{\pm 2}$, and so on, in order. In general, the physical addresses of all of the PEs in a partition of size 2^s must agree in their low order $m - s$ bit positions.

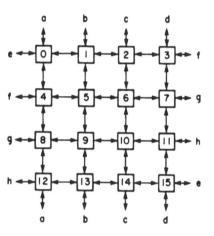

Figure 7. FNN network for N = 16. Vertical lines are $+\sqrt{N}$ and $-\sqrt{N}$. Horizontal lines are $+1$ and -1. Lower case letters indicate "wrap-around" connections.

The **Four Nearest Neighbor (FNN)** network (Figure 7) consists of four interconnection functions:

$$FNN_{+1}(P) = P + 1 \bmod N,$$
$$FNN_{-1}(P) = P - 1 \bmod N,$$
$$FNN_{+n}(P) = P + n \bmod N,$$
$$FNN_{-n}(P) = P - n \bmod N$$

where n is the square root of N and is assumed to be an integer. For example, if $N = 16$, $FNN_{+n}(2) = 6$. This network allows PE P to send data to any one of PEs $P + 1, P - 1, P + n$, or $P - n$, arithmetic mod N. The FNN network is a subset of the PM2I network. The FNN type of network was used in the Illiac IV SIMD machine [BoD72], and is included in the MPP [Bat82] and DAP [Hun81] SIMD systems. It is similar to the eight nearest neighbor network used in the CLIP4 [Duf85] machine. The "mesh" interconnection network is like the FNN network except there are no "wrap-around" connections (see Figure 7).

The operation of the FNN network in SIMD mode is similar to that for the Cube. For example, for $N = 16$, if all PEs do FNN_{+1} twice the resulting data transfer is equivalent to $PM2_{+1}$ [Sie79]. To do data transfers in MIMD mode, a sequence of FNN functions which add up to the difference between the source and destination addresses is used to transfer the data. For example, for $N = 64$, for PE 2 to transmit data to PE 28, one possible sequence would be to execute FNN_{+n} three times and FNN_{+1} twice. The route taken would be through PEs 2, 10, 18, 26, 27, and 28. As in the PM2I network, the order of performing the transfers is not important, and it is also possible to find different sets of FNN functions which would perform the same routing; e.g., going from PE 2 to PE 28 could also be accomplished by executing FNN_{-n} five times and FNN_{+1} twice.

The FNN network cannot be partitioned into independent subnetworks, each of which has the properties of a complete FNN network. In order to have a subnetwork that has the same properties as the FNN, each PE must have four interconnection functions. Allowing each PE to use all four functions, however, results in the full network.

The **Shuffle-Exchange** network (Figure 8) consists of a *shuffle* function and an *exchange* function defined by:

$$shuffle\,(p_{m-1} \cdots p_1 p_0) = p_{m-2} p_{m-3} \cdots p_1 p_0 p_{m-1}$$

$$exchange\,(p_{m-1} \cdots p_1 p_0) = p_{m-1} \cdots p_1 \bar{p}_0$$

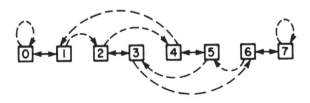

Figure 8.   Shuffle-Exchange network for $N = 8$. Solid line is exchange, dashed line is shuffle.

For example, $shuffle(6) = 5$ for $N = 8$. Shuffling a PE's address is equivalent to taking the left cyclic end-around shift of its binary representation [Sto71]. The name *shuffle* has its origin in shuffling cards by perfectly intermixing two halves of a deck, as shown in Figure 9. The exchange function is equivalent to $cube_0$. The shuffle is included in the network designs of the Omen [Hig72] and RAP [CoG74] systems. The multistage omega network is a series of m shuffles and exchanges [Law75].

The network's operation in SIMD mode is similar to that of the Cube. For example, for $N = 8$, if all PEs do two shuffles, then an exchange, and then another shuffle, the resulting data transfer is equivalent to $cube_1$ [Sie79]. In MIMD operations, data routing is accomplished by a sequence of shuffles and exchanges. For example, for $N = 16$, one way to transmit data from PE 11 (1011) to PE 0 (0000) is to per-

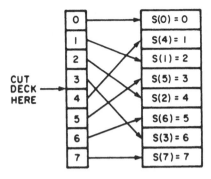

Figure 9. Perfectly shuffling a deck of eight cards. "S" stands for "shuffle."

form one exchange (1011 to 1010), a shuffle (1010 to 0101), an exchange (0101 to 0100), two shuffles (0100 to 0001), and an exchange (0001 to 0000). Even though the Shuffle-Exchange contains only two interconnection functions, it takes at most $2m - 1$ moves (at most $m - 1$ shuffles and m exchanges) to go from any source to any destination.

The Shuffle-Exchange network cannot be partitioned into independent subnetworks. The reasoning is similar to that for the FNN network.

3. The Multistage Cube Network

The multistage Cube/Shuffle-Exchange family of networks is represented in this paper by the **Generalized Cube (GC)** network [SiM81b, SiS78]. This topology is equivalent to that used by the omega [Law75], indirect binary n-cube [Pea77], STARAN flip [Bat76], multistage shuffle-exchange [ThN81], and SW-banyan $(F = S = 2)$ [GoL73] networks [SiS78]. Other networks in this family include the delta [Pat81] and baseline [WuF80]. This type of network is used or proposed for use in STARAN [Bat77], DISP [Fil82], PASM [SiS81, SiS87], Ultracomputer [GoG83], BBN Butterfly [CrG85], the Burroughs Flow Model Processor for the Numerical Aerodynamic Simulator [BaL81], the IBM RP3 [PfB85], and data flow machines [DeB80]. This network can operate in the SIMD, multiple-SIMD, MIMD, and partitionable SIMD/MIMD modes of parallelism.

A GC network with N I/O ports has $m = \log_2 N$ stages, where each stage consists of a set of N lines (*links*) connected to N/2 interchange boxes (Figure 10). Each *interchange box* is a two-input, two-output switch, and can be set to one of four legitimate states (Figure 10). The links are labeled from 0 to $N - 1$, and the labels of the links entering the upper and lower box inputs have the same labels as the upper and lower outputs, respectively. Each interchange box will be controlled independently through the use of routing tags. The "Generalized Cube" network name comes from the fact that this network is a multistage implementation of the Cube interconnection functions (defined previously) and that it was designed as a standard to represent this family of networks. Stage i of a GC topology has the capabilities of the *cube*, interconnection function because it pairs links that differ in the i-th bit position.

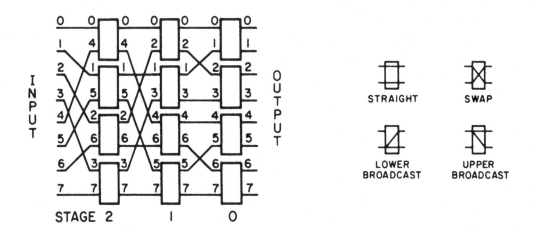

Figure 10. Generalized Cube (GC) network for $N = 8$.

In order to set the GC to perform one-to-one connections, the straight and swap box settings are used. When an interchange box in stage i is set to swap it is implementing the $cube_i$ interconnection function. For example, for $N = 8$, the path of interchange boxes from source $S = 2$ to destination $D = 4$ is set to swap in stages 2 and 1 ($cube_2$ and $cube_1$), and straight in stage 0. In general, to go from a source $S = s_{m-1} \ldots s_1 s_0$ to a destination $D = d_{m-1} \ldots d_1 d_0$, the stage i box in the path from S to D must be set to swap ($cube_i$) if $s_i \neq d_i$ and to straight if $s_i = d_i$. There is only one path from a given source to a given destination, since only stage i can determine the i-th bit of the destination address. The link label at the output of the stage i interchange box on the path from S to D is $d_{m-1} \ldots d_{i+1} d_i s_{i-1} \ldots s_1 s_0$. In order to perform a broadcast (one-to-many) connection, the lower and/or upper broadcast states of interchange boxes are also used.

Permutation connections, where each input is connected to a single distinct output, are used in SIMD mode. Since each connection in a permutation is one-to-one, only the straight and swap box settings are used. Because each of the $Nm/2$ boxes can be set to straight or swap, the GC can do $2^{Nm/2}$ out of all possible $N!$ different permutations. For $N \geq 8$, $2^{Nm/2} \ll N!$ However, the GC can do most permutations that are important in SIMD processing [Law75, Pea77].

A routing tag can be used as a header on each message to control the GC network, allowing network control to be distributed among the PEs. An m-bit tag for one-to-one (non-broadcast) connections or permutations can be computed from the source and destination address. Let $S = s_{m-1} \ldots s_1 s_0$ be the source address and $D = d_{m-1} \ldots d_1 d_0$ be the destination address. Then the tag is $T = t_{m-1} \ldots t_1 t_0 = S \oplus D$ (where "$\oplus$" means bitwise "exclusive-or") [SiM81b]; $t_i = 1$ means $s_i \neq d_i$, $t_i = 0$ means $s_i = d_i$. An interchange box in the network at stage i need only examine t_i. If $t_i = 1$, a swap is performed (i.e., $cube_i$ is performed), and if $t_i = 0$, the straight connection is used. For example, if $N = 8$,

$S = 2 = 010$, and $D = 4 = 100$, then $T = 110$, and the corresponding stage set-tings are swap, swap, and straight. Since each source generates its own tag, it is possible that a conflict will occur in the network, e.g., the tag on the upper input link of a box specifies a swap while the tag on the lower input specifies a straight. In a situation like this, one message must wait until the other has completed its transmission. Both requests cannot be serviced simultaneously. If broadcasts are allowed, then 2m bits are used in the tag [SiM81b].

GC partitioning is related to the partitioning of the single stage Cube network which was described previously. There are m ways to partition a GC of size N into two independent subnetworks of size N/2, each based on a different bit position of the I/O port addresses (i.e., a different Cube interconnection function). One way is to force all boxes in stage $m - 1$ to the straight state (i.e., disallow the use of $cube_{m-1}$). This would form two subnetworks, one consisting of those I/O ports with a 0 in the high order bit position of their addresses and the other consisting of those ports with a 1 in the high order bit position. These two groups could com-municate with each other only by using the swap setting in stage $m - 1$ (i.e., the $cube_{m-1}$ function). By forcing this stage to straight, the subnetworks are inde-pendent and have full use of the rest of the network (stages $m - 2$ to 0, corre-sponding to $cube_{m-2}$ to $cube_0$) (Figure 11). Since each subnetwork is a GC, each can be further subdivided. The only constraints are that the size of each subnet-work must be a power of two, the physical addresses of the I/O ports of a subnet-work of size 2^s must all agree in any fixed set of $m - s$ bit positions, the network stages used by this subnetwork corresponding to these $m-s$ bit positions must be set to straight, and each I/O port can belong to at most one subnetwork.

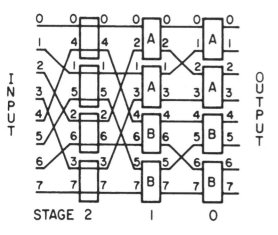

STAGE 2   I   O

Figure 11. Partitioning the GC for N = 8 into two subnetworks of size four (A and B) based on the high order bit position.

An aspect of all multistage networks is the way in which paths through the network are established and released. *Circuit switching* implies the establishing of a com-plete path between the source and the destination. Once such a path is estab-lished, subsequent transfers through the path move the data from a network input port directly to a network output port. In a *packet-switched* network, often referred to as *message-switched,* data and routing information are collected into a packet, and this packet makes its way from one stage to the next stage, releasing links and interchange boxes immediately after using them. Thus, each interchange box is capable of storing packets and a packet uses only one interchange box at a time. The tradeoffs between these two approaches are currently under study [DaS86].

4. The Extra Stage Cube Network

In the GC network there is only one path from a given network input to a given output, so the network is not single fault tolerant. Techniques such as test patterns [DaH85, FeW81, FeZ85] or dynamic parity checking [SiM81b] for fault detection and location have been described for use with the GC topology. Different approaches to fault tolerant GC-type networks have been studied [AdA87]. This section discusses one way to recover once a fault is located. The **Extra Stage Cube (ESC)** network is a single fault-tolerant network derived from the GC network, capable of operating in SIMD, multiple-SIMD, MIMD, and partitionable SIMD/MIMD environments [AdS82, AdS84].

The ESC network is formed from the GC by adding

1. one extra stage at the input and hardware to allow the bypass, when desired, of the extra stage (stage m) or the output stage (stage 0), and
2. using dual I/O lines to the PEs (Figure 12).

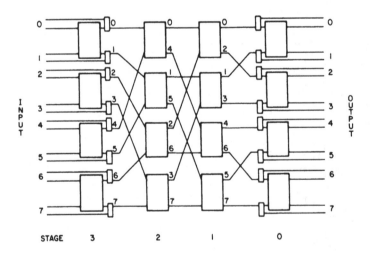

Figure 12.   The Extra State Cube (ESC) network for N = 8.

Both stages m and 0 implement the $cube_0$ interconnection function. Enabling and disabling stages m and 0 may be performed by a system control unit (Figure 13).

In normal operation, stage m is disabled (bypassed) and stage 0 is enabled, resulting in a topology identical to a GC. If a fault is found, the network is reconfigured. If the fault is in stage 0, then stage m is enabled and stage 0 is disabled,

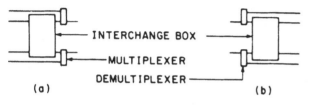

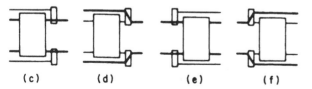

Figure 13.   Bypass hardware in the ESC.
(a) Details of stage m interchange box; (b) Details of stage 0 interchange box; (c) Stage m enabled; (d) Stage m disabled; (e) Stage 0 enabled; (f) Stage 0 disabled.

and stage m handles the $cube_0$ function instead of stage 0. For a fault in a link or a box in stages $m - 1$ to 1, both stages m and 0 will be enabled. Stage m of the ESC network allows access to two distinct stage $m - 1$ inputs, S and $cube_0(S)$. Stages $m - 1$ to 0 of the ESC network form a GC topology, so each of the two stage $m - 1$ inputs has a single path to the destination and these paths are distinct except for the stage m and 0 boxes, which are fault-free in this case. The ESC network path that uses the straight setting in stage m is called the *primary path* and the path that uses the swap setting is called the *secondary path*. Thus, after a single fault at least one fault-free path must exist. This is demonstrated for the paths from input 6 to output 3 in Figure 14. The primary path (solid line) from 6 to 3 which enters stage 2 in link 6 goes through links with labels that agree with $S = 6$ in the 0-th bit position (it is 0). The secondary path (dotted line) from 6 to 3 which enters stage 2 on link $cube_0(6) = 7$ goes through links with labels that agree with $cube_0(6) = 7$ in the 0-th bit position (it is 1). Broadcasting is done similarly. SIMD permutations performable by the GC will, in general, require two passes through the ESC if a fault has occurred.

If an input line connecting a PE to a stage m multiplexer fails, stage m is enabled and forced to straight. Thus, the non-faulty input line will be used. If the fault is on an input line to a stage m interchange box, that line is currently unused and the system continues to ignore the faulty line. If an output line from a stage 0 box to a PE is faulty, the network is reconfigured as if stage 0 is faulty. If the fault is on an output line from a demultiplexer to a PE, that line is currently unused and the system continues to ignore the faulty line.

The ESC uses an $(m + 1)$-bit routing tag $T = t'_m \ldots t'_1 t'_0$ for one-to-one connections (for broadcast connections $2m + 2$ bits are used). Stage i interprets t'_i like with the GC. If the network is fault-free, stage m is disabled, and the m-th bit will be ignored, and the routing tag is given by $T = X t_{m-1} \ldots t_1 t_0$, where $t_{m-1} \ldots t_1 t_0 = T$, the tag used in the GC, and "X" can be 0 or 1 ("don't care"). If there is a fault in a network link or box in stages $m - 1$ to 1, stage m is enabled. The primary path is

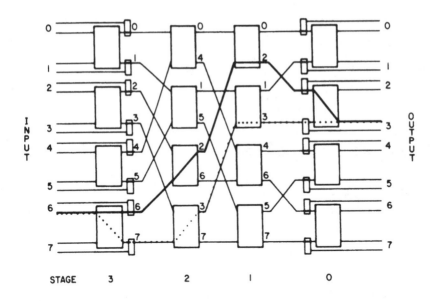

Figure 14. The paths from input 6 to output 3 in the ESC network for N = 8, when both stages 3 and 0 are enabled.

used if it is not faulty; otherwise, the secondary path is used. The tag $T = t_{m-1} \dots t_1 t_0$ yields the primary path, and $T = 1 t_{m-1} \dots t_1 \overline{t_0}$ the secondary path (stage 0 uses $\overline{t_0}$ instead of t_0 to compensate for the swap ($cube_0$) already performed by stage m). For Figure 14, $T = 0101$ specifies the primary path (solid line), while $T = 1100$ specifies the secondary path (dotted line). If the fault is in stage 0, stage m is enabled and stage 0 disabled, and the tag $T = t_0 t_{m-1} \dots t_1 X$ (setting $t_m' = t_0$) is used.

The following method is used to determine if a primary path is faulty. If a fault is located in stages $m - 1$ to 1, or any link, each PE receives a *fault label*. If the link labeled j between stages i and $i - 1$ fails, each PE receives the label (i, j). If the stage i box with outputs j and k fails, each PE receives the label (i, j, k). A source S forms $d_{m-1} \dots d_{i+1} d_i s_{i-1} \dots s_1 s_0$, the stage i link label on the primary path from S to D. If this matches j (or k) in the fault label, the primary path is faulty; otherwise, it is fault-free.

The ESC network can be partitioned in ways similar to the GC, with each subnetwork retaining all the properties of an ESC network. For example, partitioning based on bit $m - 1$ is shown in Figure 15. Partitioning cannot be based on stages m or 0 (bit 0 of the physical port addresses) because these two stages are needed to implement $cube_0$ to provide the redundant paths. The other aspects of partitioning are the same as for the GC, e.g., subdividing subnetworks. If the stage m and 0 box bypassing is controlled for each box individually, rather than by using a single control signal for the entire stage, then each partition can bypass stages independently and each subnetwork will be single fault tolerant.

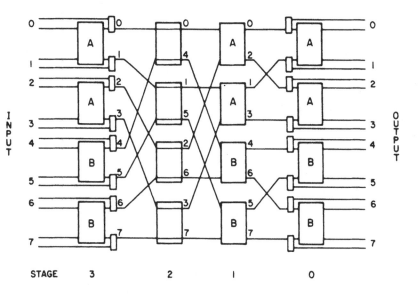

STAGE   3   2   I   0

Figure 15.   The ESC network for $N = 8$ partitioned into two independent subnetworks of size four, based on the high order bit position. The labels A and B denote the two subnetworks.

5. Data Manipulator Networks

"Data manipulator" networks are based on the PM2I interconnection functions defined in Section 2. The class of data manipulator networks includes the data manipulator [Fen74], the Augmented Data Manipulator (ADM) [SiS78], the Inverse Augmented Data Manipulator (IADM) [SiM81a], and the gamma [PaR82] multistage interconnection networks. The ADM network is used as a representative of this class. It can operate in the SIMD, multiple-SIMD, MIMD, and partitionable SIMD/MIMD modes of parallelism.

The **data manipulator** network [Fen74] (Figure 16) has m stages, each *stage* consisting of N switching elements (*switches*) and three output links per switch. There is also an $(m + 1)$-st column of network output switches. The stages are ordered from $m - 1$ to 0, and at stage i of the network switch j is connected to switch $j - 2^i$ mod N of stage $i - 1$ (i.e., $PM2_{-i}$), switch j of stage $i - 1$ (i.e., straight), and switch $j + 2^i$ mod N of stage $i - 1$ (i.e., $PM2_{+i}$). Each switch selects one of its input links and connects it either to one of its output links (a one-to-one setting) or to two or three of its output links (a broadcast setting). The **Augmented Data Manipulator (ADM)** network is a data manipulator network where each switch can be set independently [McS82, SiS78, SiM81a]. To establish a path from a source S of the ADM network to a destination D, links whose "sum" mod N is $D - S$ mod N are used, e.g., for $N = 8$, a path from $S = 1$ to $D = 6$ is $+2^2$ in stage 2 and $+2^0$ in stage 0 (the straight connection in stage 1 contributes nothing to the sum). Relat-

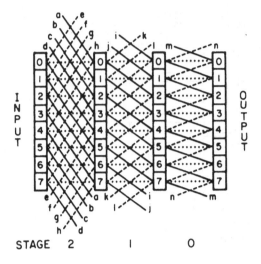

STAGE 2 1 0

Figure 16. The data manipulator network for N = 8. Straight connections are shown by the dotted lines, PM2₊, by the solid lines, and PM2₋, by the dashed lines. The lower case letters represent "end-around" connections.

ing this to the single stage PM2I network, $PM2_{+0}(PM2_{+2}(1)) = 6$. Other paths between 1 and 6 exist: $+2^2$, $+2^1$, -2^0; and straight, -2^1, -2^0. The existence of multiple (not necessarily disjoint) paths between source/destination pairs S and D ($S \neq D$) gives the ADM greater flexibility than the GC. This is different from the multiple path feature of the ESC, which, with stages m and 0 enabled, has exactly two disjoint paths between any source/destination pair.

The set of permutations performable by the ADM network is a superset of that of the GC, and, for many permutations, there exist multiple network settings which result in the same permutation. Details on the number of permutations performable by the ADM network can be found in [Lei85].

To specify an arbitrary path in an ADM network, a *full routing tag*, $F = f_{2m-1} \ldots f_0$, can be used. A stage i switch examines f_i and f_{m+i}. If $f_i = 0$, the straight link is used. If $f_i = 1$, f_{m+i} is examined. If $f_{m+i} = 0$, the $+2^i$ link is used. If $f_{m+i} = 1$, the -2^i link is used. For example, for $N = 16$, the tag $F = 00111011$ will go from 1 to 6 on the path $+2^3$, straight, -2^1, -2^0. One way to calculate a full tag F from source S to destination D is to set bits 0 to $m-1$ of F to $S \oplus D$ and bits m to $2m-1$ to $S\bar{D}$, e.g., for $S = 1$ and $D = 6$, $F = 00010111$ (the path $+2^2$, $+2^1$, -2^0). A *natural routing tag* uses only one bit for the sign. All the non-straight links traversed are of the same sign. An $m+1$ bit routing tag is formed by computing the *signed magnitude difference* between the destination and the source: $T = t_m \ldots t_0 = D - S$. The *sign bit* is t_m, where $t_m = 0$ indicates positive or zero (i.e., $D > S$), and $t_m = 1$ indicates negative (i.e., $D < S$). Bits $t_{m-1} \ldots t_0$ equal the absolute value of D−S, the *magnitude* of the difference. The natural routing tag is interpreted in the same way as the full routing tag, except t_m is used as the sign bit at every stage. For example, if $N = 16$, $S = 13$, and $D = 6$, then $T = -7 = 10111$, and the path is straight, -2^2, -2^1, -2^0. For any natural tag T for S to D (S≠D) an equivalent routing tag from S to D can be computed that uses links of the opposite sign by taking the two's complement of T. This idea is used in [McS82] to develop techniques for routing around faulty or busy switches.

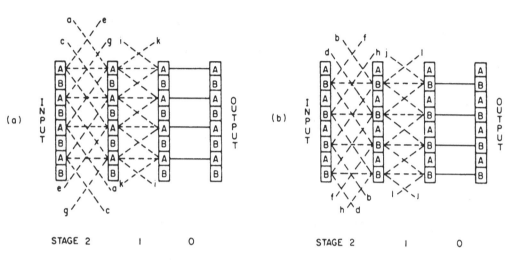

Figure 17. Partitioning a size-eight ADM network into two independent size four networks.
(a) The A subnetwork, I/O ports 0, 2, 4, and 6. (b) The B subnetwork, I/O ports 1, 3, 5, and 7.

The partitioning of the ADM is related to the partitioning of the PM2I single stage network (Section 2). Consider partitioning an ADM network of size N into two independent ADM subnetworks, each of size N/2. The only way to do this is to treat the even numbered I/O ports as one subnetwork and the odd numbered ports as the other (Figure 17). By setting all of the stage 0 connections to straight, no PE attached to one subnetwork can communicate with a PE attached to the other. This is because the only way a PE with a 0 in the 0-th bit of its address can route data to a PE with a 1 in the 0-th bit of its address is by using a $+2^0$ or -2^0 connection at stage 0. Each subnetwork can be independently further subdivided. The only constraints are that the size of each subnetwork must be a power of two, the physical addresses of the I/O ports of a subnetwork of size 2^s must agree in their low order $m - s$ bit positions, and the switches used by this subnetwork are set to straight in stages 0 to $(m - s) - 1$.

6. Dynamic Redundancy Network

For systems which contain a large number of PEs, the possibility of PE failures cannot be neglected. One way to tolerate PE failures is the use of *dynamic redundancy*, i.e., spare PEs are used to replace faulty PEs when a fault is detected. The

Dynamic Redundancy (DR) network is a fault-tolerant GC network which can be used directly to support systems containing spare PEs [JeS86a, JeS86b, JeS87]. The DR network can tolerate any single switch or link failure (including input and output stage switches and the links between PEs and I/O ports of the network) and provide the necessary capabilities for the system to recover from any single PE failure. DR networks can be used to support SIMD, multiple-SIMD, MIMD, and partitionable SIMD/MIMD systems. As in the ESC section, the concern here is with fault recovery after fault detection and location.

The design of the DR network is based on the interconnection graph of the GC network. Figure 18 shows an equivalent SW-banyan graph of the GC network with $N = 8$ [MaM81, Sie85]. Consider each vertex as a switch and each edge as a link. Using this representation of the GC, each stage has N switches and 2N links. Recall that the basic concept underlying the GC topology is that at stage i PEs whose network input port labels differ in the i-th bit position can swap data (i.e., stage i implements $cube_i$). Using the representation in Figure 10, at stage i, link j and link $cube_i(j)$ can swap data. Using the representation in Figure 18, at stage i, switch j and switch $cube_i(j)$ can swap data; i.e., switch j in stage i is connected to switch $cube_i(j)$ in stage $i - 1$, $0 \leq j < N$, $0 \leq i < m$.

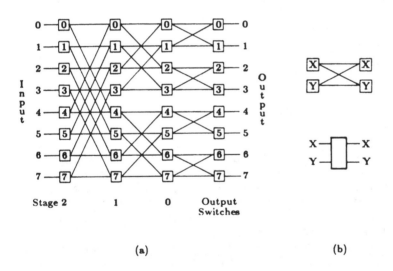

Figure 18. SW-banyan graph of the GC network. (a) Graphical interpretation of the GC for $N = 8$. (b) Relationship between graphical interpretation and interchange box representation.

Since the DR interconnection network is to be used in systems which contain N functioning PEs and σ spare PEs, it must contain $N + \sigma$ I/O ports. Like the GC, the DR network contains m stages, where $N = 2^m$ (Figure 19). Stages are ordered from $m - 1$ to 0 and each stage has $N + \sigma$ switches followed by $3(N + \sigma)$ links. In addition, there are $N + \sigma$ output switches. Let the PEs of the system be numbered from 0 to $N + \sigma - 1$. PE j of the system is connected to the input of switch j of stage $m - 1$ and to the network output switch j, where $0 \leq j \leq N + \sigma - 1$. Switch j at stage i is connected to switch $(j - 2^i) \bmod (N + \sigma)$ of stage $i - 1$, to switch j of stage $i - 1$, and to switch $(j + 2^i) \bmod (N + \sigma)$ of stage $i - 1$. The DR network topology is similar to the ADM network topology, except that the ADM network has N I/O ports and uses mod N instead of mod $(N+\sigma)$ in the connection functions. Because of the spare I/O ports, the interconnection and fault-tolerance capabilities of the DR network are quite different from those of the ADM.

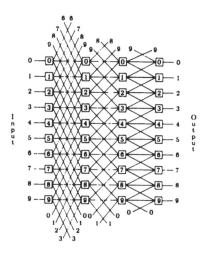

Figure 19. A Dynamic Redundancy (DR) network for $N=8$ and $\sigma=2$.

Since only N PEs are functioning at a time, only N of the $N+\sigma$ network I/O ports are needed at a time to provide the inter-PE communication for the system. When there are no faults, PEs 0 to $N-1$ are used as the functioning PEs, and switches numbered from 0 to $N-1$ and their associated links emulate the GC, which is a subgraph of the DR. At stage i, DR switches j and cube$_i$(j) emulate the stage i interchange box with inputs j and cube$_i$(j); assuming $j <$ cube$_i$(j), setting switch j to $+2^i$ and switch cube$_i$(j) to -2^i emulates "swap," and setting switches j and cube$_i$(j) to straight emulates "straight." (Upper and lower broadcast are defined similarly.)

If PE j or switch j (any stage) or a link attached to switch j (any stage) is found to be faulty, the system is reconfigured so that the PE and switches physically numbered P are logically numbered $P - (j + \sigma) \bmod (N + \sigma)$. Then PEs logically numbered 0 to $N-1$ are used as the functioning PEs, and switches logically numbered 0 to $N-1$ and their associated links emulate the GC, in the same manner as described for the fault-free case above (but based on the logical numbering). All GC performable permutations are performable in a single pass through the DR even after a fault occurs. An example of reconfiguration is shown in Figure 20, where each logical number can be obtained by subtracting 9 (mod 10) from the corresponding physical number specified.

Routing tags for the DR are based on the logical numbers for the PEs and switches, where for the no-fault situation physical PE i and switch i are also logically numbered i, $0 \leq i < N$. Given a source PE with logical address S and a destination PE with logical address D, then the *exclusive-or routing tag* T can be derived by taking the bitwise exclusive-or of S and D (as done for the GC). When a switch

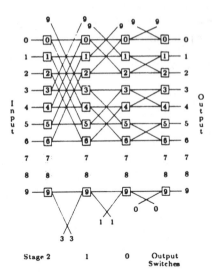

Figure 20. Reconfiguration of a DR network with N = 8 and σ = 2 when PE 7 is faulty. The solid lines show the GC subgraph of the DR network.

in stage i receives a message, it examines t_i. If $t_i = 0$, then the straight link is used. If $t_i = 1$, then a switch has to decide to use the $+2^i$ link or the -2^i link. Let $W = w_{m-1} \ldots w_1 w_0$ represent the logical address of the switch. If $w_i = 0$, switch W will use $+2^i$ link, otherwise it will use -2^i link. This will emulate a stage i swap, i.e., connect W to cube$_i$(W). This routing tag scheme necessitates adding a one-bit flag in each switch to store the i-th bit of the logical address of the switch, w_i. These flags can be set by special messages sent from PEs during system initialization and after each reconfiguration due to a fault. This allows the same routing tags used in the GC network to be used directly to control the DR network. An alternative method, which does not require each switch to store its logical number, sends the logical address of the source PE, an additional m bits, in the tag. A connection path from a logical source PE S to a logical destination PE D will use the switch with logical address $d_{m-1} \ldots d_{i+1} s_i \ldots s_0$ at stage i (this was mentioned for the GC in Section 3). Thus, $w_i = s_i$, and each switch can use s_i in place of w_i. (Broadcasting on the DR is an extension of the one-to-one tag schemes.)

The partitionability of the DR network depends on the value of σ. When σ is even, the DR network can be partitioned into two independent subnetworks by setting all switches in stage 0 to straight. The theory underlying this is similar to that for partitioning the ADM network. In general, each subnetwork of size N/2 + σ/2 is a single fault tolerant DR network and can be partitioned. The limit is that a DR network of size N + σ, where N = 2^m and σ = 2^q, can be partitioned into σ independent, single fault tolerant subnetworks of size N′ + σ′, where N′ = 2^{m-q} and σ′ = 1. The physical addresses of all the switches in a subnetwork of size $2^{m-r} + 2^{q-r}$ must agree in their lower-order r bit positions, and partitions can be of different sizes.

7. Summary

This paper overviewed an important set of interconnection networks. In Section 2, four basic interconnection networks, the Cube, FNN, PM2I, and Shuffle-Exchange, were defined, and single stage implementations of these networks were examined. The use of these networks in SIMD and MIMD environments was discussed, as was their partitionability.

The GC (a representative of the multistage Cube/Shuffle-Exchange family of networks), ADM (a representative of the multistage PM2I (data manipulator) family of networks), ESC, and DR multistage networks were described in Sections 3 through 6. Advantages of all four multistage networks described include: hardware complexity proportional to $Nlog_2N$, distributed control using routing tags, the capability for one PE to broadcast to all or a subset of the others, partitionability into independent subnetworks, adaptability for use in SIMD or MIMD mode, and options for a variety of implementation methods. The ADM is more flexible than the GC, but also more costly to implement [McS85]. The ESC and DR networks are single fault tolerant variations of the GC. The DR network supports the use of spare PEs and can do GC permutations in a single pass after a single fault, but may require the reloading of PE memories after reconfiguration for a fault (due to the new logical PE numbers). The ESC network is less costly and does not require PE reloading after a fault, but it requires two passes, in general, to do a GC permutation after a single fault and does not support the use of spare PEs.

The purpose of this paper was to introduce some communication techniques used in large-scale parallel processing systems. Further information and pointers to additional reading are in [Sie85] and in the other references cited.

Acknowledgments — Figure 1 to Figure 11, Figure 16, and Figure 17 are from [Sie85]. All figures were drawn by Sharon Katz.

8. References

[AdA87]   G. B. Adams III, D. P. Agrawal, and H. J. Siegel
          "A survey and comparison of fault-tolerant multistage interconnection networks,"
          Computer, Vol. 20, June 1987, pp. 14-27.

[AdS82]   G. B. Adams III and H. J. Siegel
          "The extra stage cube: a fault-tolerant interconnection network for supersystems,"
          IEEE Trans. Computers, Vol. C-31, May 1982, pp. 443-454.

[AdS84]   G. B. Adams III and H. J. Siegel
          "Modifications to improve the fault tolerance of the extra stage cube interconnection network,"
          1984 International Conf. Parallel Processing, Aug. 1984, pp. 169-173.

[BaB68]   G. H. Barnes, R. Brown, M. Kato, D. J. Kuck, D. L. Slotnick, and R. A. Stokes
          "The Illiac IV computer,"
          IEEE Trans. Computers, Vol. C-17, Aug. 1968, pp. 746-757.

[Bae80]   J-L. Baer
"Computer Systems Architecture,"
Computer Science Press, Potomac, MD, 1980.

[BaL81]   G. H. Barnes and S. F. Lundstrom
"Design and validation of a connection network for many-processor multi-processor systems,"
Computer, Vol. 14, Dec. 1981, pp. 31-41.

[Bat76]   K. E. Batcher
"The flip network in STARAN,"
1976 International Conf. Parallel Processing, Aug. 1976, pp. 65-71.

[Bat77]   K. E. Batcher
"STARAN series E,"
1977 International Conf. Parallel Processing, Aug. 1977, pp. 140-143.

[Bat82]   K. E. Batcher
"Bit serial parallel processing systems,"
IEEE Trans. Computers, Vol. C-31, May 1982, pp. 377-384.

[BeD85]   J. Beetem, M. Denneau, and D. Weingarten
"The GF11 supercomputer,"
12th Ann. International Symp. Computer Architecture, June 1985, pp. 108-115.

[BoD72]   W. J. Bouknight, S. A. Denenberg, D. E. McIntryre, J. M. Randall, A. H. Sameh, and D. L. Slotnick
"The Illiac IV system,"
Proc. IEEE, Vol. 60, Apr. 1972, pp. 369-388.

[BrH83]   G. Broomell and J. R. Heath
"Classification categories and historical development of circuit switching topologies,"
ACM Computing Surveys, Vol. 15, June 1983, pp. 95-133.

[CoG74]   G. R. Couranz, M. S. Gerhardt, and C. J. Young
"Programmable RADAR signal processing using the RAP,"
1974 Sagamore Computer Conf. Parallel Processing, Aug. 1974, pp. 37-52.

[CrG85]   W. Crowther, J. Goodhue, E. Starr, R. Thomas, W. Williken, and T. Blackadar
"Performance measurements on a 128-switch Butterfly parallel processor,"
1985 International Conf. Parallel Processing, Aug. 1985, pp. 531-540.

[DaH85]   N. J. Davis IV, W. T.-Y. Hsu, and H. J. Siegel
"Fault location techniques for distributed control interconnection networks,"
IEEE Trans. Computers, Vol. C-34, Oct. 1985, pp. 902-910.

[DaS86]   N. J. Davis IV and H. J. Siegel
"Performance analysis of multiple-packet multistage 'cube' networks and comparison to circuit switching,"
1986 International Conf. Parallel Processing, Aug. 1986, pp. 108-114.

[DeB80]   J. B. Dennis, G. A. Boughton, and C. K. C. Leung
"Building blocks for data flow prototypes,"
7th Ann. Symp. Computer Architecture, May 1980, pp. 1-8.

[Duf85]   M.J.B. Duff
"Real Applications on CLIP4,"
in Integrated Technology for Parallel Image Processing, S. Levialdi, ed., Academic Press, Orlando, Florida, 1985, pp. 153-165.

- 57 -

[Fen74]   T. Y. Feng
            "Data manipulating functions in parallel processors and their implementa-
            tions,"
            IEEE Trans. Computers, Vol. C-23, Mar. 1974, pp. 309-318.

[Fen81]   T. Y. Feng
            "A survey of interconnection networks,"
            Computer, Vol. 14, Dec. 1981, pp. 12-27.

[FeW81]   T. Y. Feng and C-L. Wu
            "Fault-diagnosis for a class of multistage interconnection networks,"
            IEEE Trans. Computers, Vol. C-30, Oct. 1981, pp. 743-758.

[FeZ85]   T. Y. Feng and Q. Zhang
            "Fault diagnosis of multistage interconnection networks with four valid
            states,"
            5th International Conf. Distributed Computing Systems, May 1985, pp. 218-
            226.

[Fil82]   A. E. Filip
            "A distributed signal processing architecture,"
            3rd International Conf. Distributed Computing Systems, Oct. 1982, pp. 49-55.

[Fly66]   M. J. Flynn
            "Very high-speed computing systems,"
            Proc. IEEE, Vol. 54, Dec. 1966, pp. 1901-1909.

[GoG83]   A. Gottlieb, R. Grishman, C. P. Kruskal, K. P. McAuliffe, L. Rudolph, and M.
            Snir
            "The NYU Ultracomputer -- designing an MIMD shared-memory parallel
            computer,"
            IEEE Trans. Computers, Vol. C-32, Feb. 1983, pp. 175-189.

[GoL73]   G. R. Goke and G. J. Lipovski
            "Banyan networks for partitioning multiprocessor systems,"
            1st Ann. Symp. Computer Architecture, Dec. 1973, pp. 21-28.

[HaD85]   M. T. Hagan, H. B. Demuth, and P. H. Singgih
            "Parallel signal processing on the HEP,"
            1985 International Conf. on Parallel Processing, Aug. 1985, pp. 599-606.

[Hig72]   L. C. Higbie
            "The Omen computer: associative array processor,"
            IEEE Computer Society Compcon 72, Sept. 1972, pp. 287-290.

[Hil85]   W. D. Hillis
            "The Connection Machine,"
            The M.I.T. Press, Cambridge, MA, 1985.

[HoJ81]   R. W. Hockney and C. R. Jeshope
            "Parallel Computers,"
            Adam Hilger Ltd., Bristol, England, 1981.

[Hun81]   D. J. Hunt
            "The ICL DAP and its application to image processing,"
            in Languages and Architectures for Image Processing, M. J. B. Duff and S.
            Levialdi, eds., Academic Press, London, England, 1981, pp. 275-282.

[HwB84]   K. Hwang and F. Briggs
            "Computer Architecture and Parallel Processing,"
            McGraw-Hill, New York, NY, 1984.

[Int85]   Intel Corporation
             "A New Direction in Scientific Computing,"
             Order # 28009-001, Intel Corporation, 1985.

[JeS86a]   M. Jeng and H. J. Siegel
             "A fault-tolerant multistage interconnection network for multiprocessor sys-
             tems using dynamic redundancy,"
             6th International Conf. Distributed Computing Systems, June 1986, pp. 70-
             77.

[JeS86b]   M. Jeng and H. J. Siegel
             "Implementation approach and reliability estimation of dynamic redundancy
             networks,"
             1986 Real-Time Systems Symp., Dec. 1986, pp. 79-81.

[JeS87]   M. Jeng and H. J. Siegel
             "Design and analysis of dynamic redundancy networks"
             IEEE Trans. Computers, to appear

[Kuc78]   D. J. Kuck
             "The Structure of Computers and Computations, Vol. 1,"
             John Wiley and Sons, New York, NY, 1978.

[Law75]   D. H. Lawrie
             "Access and alignment of data in an array processor,"
             IEEE Trans. Computers, Vol. C-24, Dec. 1975, pp. 1145-1155.

[Lel85]   M. D. P. Leland
             "On the power of the augmented data manipulator network,"
             1985 International Conf. Parallel Processing, Aug. 1985, pp. 74-78.

[MaG79]   G. M. Masson, G. C. Gingher, and S. Nakamura
             "A sampler of circuit switching networks,"
             Computer, Vol. 12, June 1979, pp. 32-48.

[MaM81]   M. Malek and W. W. Myre
             "A description method of interconnection networks,"
             IEEE Technical Committee on Distributed Processing Quarterly, Feb. 1981,
             pp. 1-6.

[McS82]   R. J. McMillen and H. J. Siegel
             "Routing schemes for the augmented data manipulator network in an MIMD
             system,"
             IEEE Trans. Computers, Vol. C-31, Dec. 1982, pp. 1202-1214.

[McS85]   R. J. McMillen and H. J. Siegel
             "Evaluation of cube and data manipulator networks,"
             Journ. Parallel and Distributed Computing, Vol. 2, Feb. 1985, pp. 79-107.

[Nut77]   G. J. Nutt
             "Microprocessor implementation of a parallel processor,"
             4th Ann. Symp. Computer Architecture, Mar. 1977, pp. 147-152.

[OkT82]   Y. Okada, H. Tajima, and R. Mori
             "A reconfigurable parallel processor with microprogram control,"
             IEEE Micro, Vol. 2, Nov. 1982, pp. 48-60.

[PaR82]   D. S. Parker and C. S. Raghavendra
             "The gamma network: a multiprocessor interconnection network with
             redundant paths,"
             9th Ann. Symp. Computer Architecture, Apr. 1982, pp. 73-80.

[Pat81]   J. H. Patel
            "Performance of processor-memory interconnections for multiprocessors,"
            IEEE Trans. Computers, Vol. C-30, Oct. 1981, pp. 771-780.

[Pea77]   M. C. Pease III
            "The indirect binary n-cube microprocessor array,"
            IEEE Trans. Computers, Vol. C-26, May 1977, pp. 458-473.

[PfB85]   G. F. Pfister, W. C. Brantley, D. A. George, S. L. Harvey, W. J. Kleinfelder,
            K. P. McAuliffe, E. A. Melton, V. A. Norton, and J. Weiss
            "The IBM Research Parallel Processor Prototype (RP3): introduction and
            architecture,"
            1985 International Conf. Parallel Processing, Aug. 1985, pp. 764-771.

[Sei85]   C. L. Seitz
            "The Cosmic Cube,"
            Comm. ACM, Jan. 1985, pp. 22-33.

[SeU80]   M. C. Sejnowski, E. T. Upchurch, R. N. Kapur, D. P. S. Charlu, and G. J.
            Lipovski
            "An overview of the Texas Reconfigurable Array Computer,"
            AFIPS 1980 Nat'l Computer Conf., June 1980, pp. 631-641.

[Sie77]   H. J. Siegel
            "Analysis techniques for SIMD machine interconnection networks and the
            effects of processor address masks,"
            IEEE Trans. Computers, Vol. C-26, Feb. 1977, pp. 153-161.

[Sie79]   H. J. Siegel
            "A model of SIMD machines and a comparison of various interconnection
            networks,"
            IEEE Trans. Computers, Vol. C-28, Dec. 1979, pp. 907-917.

[Sie80]   H. J. Siegel
            "The theory underlying the partitioning of permutation networks,"
            IEEE Trans. Computers, Vol. C-29, Sept. 1980, pp. 791-801.

[Sie85]   H. J. Siegel
            "Interconnection Networks for Large-Scale Parallel Processing: Theory and
            Case Studies,"
            Lexington Books, D.C. Heath and Company, Lexington, MA, 1985.

[SiM79]   H. J. Siegel, R. J. McMillen, and P. T. Mueller, Jr.
            "A survey of interconnection methods for reconfigurable parallel processing
            systems,"
            AFIPS 1979 Nat'l Computer Conf., June 1979, pp. 529-542.

[SiM81a]   H. J. Siegel and R. J. McMillen
            "Using the augmented data manipulator network in PASM,"
            Computer, Vol. 14, Feb. 1981, pp. 25-33.

[SiM81b]   H. J. Siegel and R. J. McMillen
            "The multistage cube: a versatile interconnection network,"
            Computer, Vol. 14, Dec. 1981, pp. 65-76.

[SiS78]   H. J. Siegel and S. D. Smith
            "Study of multistage SIMD interconnection networks,"
            5th Ann. Symp. Computer Architecture, Apr. 1978, pp. 223-229.

[SiS81]   H. J. Siegel, L. J. Siegel, F. C. Kemmerer, P. T. Mueller, Jr., H. E. Smalley, Jr., and S. D. Smith
"PASM: a partitionable SIMD/MIMD system for image processing and pattern recognition,"
IEEE Trans. Computers, Vol. C-30, Dec. 1981, pp. 934-947.

[SiS87]   H. J. Siegel, T. Schwederski, J. T. Kuehn, and N. J. Davis IV
"An overview of the PASM parallel processing system,"
in Computer Architecture, D. D. Gajski, V. M. Milutinovic, H. J. Siegel, and B. P. Furht, eds., IEEE Computer Society Press, Washington, DC, 1987, pp. 387-407.

[Sto71]   H. S. Stone
"Parallel processing with the perfect shuffle,"
IEEE Trans. Computers, Vol. C-20, Feb. 1971, pp. 153-161.

[SuB77]   H. Sullivan, T. R. Bashkow, and K. Klappholz
"A large-scale homogeneous, fully distributed parallel machine,"
4th Ann. Symp. Computer Architecture, Mar. 1977, pp. 105-124.

[SwF77]   R. J. Swan, S. Fuller, and D. P. Siewiorek
"Cm*: a modular multimicroprocessor,"
AFIPS 1977 Nat'l Computer Conf., June 1977, pp. 637-644.

[ThM79]   K. J. Thurber and G. M. Masson
"Distributed-Processor Communication Architecture,"
Lexington Books, D. C. Heath and Company, Lexington, MA, 1979.

[ThN81]   S. Thanawastien and V. P. Nelson
"Interference analysis of shuffle/exchange networks,"
IEEE Trans. Computers, Vol. C-30, August 1981, pp. 545-556.

[Thu79]   K. J. Thurber
"Parallel processor architectures - part 1: general purpose systems,"
Computer Design, Vol. 18, Jan. 1979, pp. 89-97.

[Wes72]   A. H. Wester
"Special features in SIMDA,"
1972 Sagamore Computer Conf. Parallel Processing, Aug. 1972, pp. 29-40.

[WuF80]   C-L. Wu and T. Y. Feng
"On a class of multistage interconnection networks,"
IEEE Trans. Computers, Vol. C-29, Aug. 1980, pp. 694-702.

[WuF84]   C.-L. Wu and T.Y. Feng, eds.
"Tutorial: Interconnection Networks for Parallel and Distributed Processing,"
IEEE Computer Society Press, Silver Spring, MD, 1984.

Two Fundamental Issues in Multiprocessing

Arvind
Robert A. Iannucci

Laboratory for Computer Science
Massachusetts Institute of Technology
Cambridge, Massachusetts 02139 - USA

Abstract

A general purpose multiprocessor should be scalable, *i.e.* show higher perform-ance when more hardware resources are added to the machine. Architects of such multiprocessors must address the loss in processor efficiency due to two funda-mental issues: long memory latencies and waits due to synchronization events. It is argued that a well designed processor can overcome these losses provided there is sufficient parallelism in the program being executed. The detrimental effect of long latency can be reduced by instruction pipelining, however, the restriction of a single thread of computation in von Neumann processors severely limits their ability to have more than a few instructions in the pipeline. Further-more, techniques to reduce the memory latency tend to increase the cost of task switching. The cost of synchronization events in von Neumann machines makes decomposing a program into very small tasks counter-productive. Dataflow machines, on the other hand, treat each instruction as a task, and by paying a small synchronization cost for each instruction executed, offer the ultimate flexi-bility in scheduling instructions to reduce processor idle time.

Key words and phrases: caches, cache coherence, dataflow architectures, hazard resolution, instruction pipelining, LOAD/STORE architectures, memory latency, multiprocessors, multi-thread architectures, semaphores, synchronization, von Neumann architecture.

1. Importance of Processor Architecture

Parallel machines having up to several dozen processors are commercially avail-able now. Most of the designs are based on von Neumann processors operating out of a shared memory. The differences in the architectures of these machines in terms of processor speed, memory organization and communication systems, are significant, but they all use relatively conventional von Neumann processors. These machines represent the general belief that processor architecture is of little importance in designing parallel machines. We will show the fallacy of this

This report describes research done at the Laboratory for Computer Science of the Massa-chusetts Institute of Technology. Funding for the Laboratory is provided in part by the Advanced Research Projects Agency of the Department of Defense under Office of Naval Research contracts N00014-83-K-0125 and N00014-84-K-0099. The second author is employed by the International Business Machines Corporation.

assumption on the basis of two issues: *memory latency* and *synchronization*. Our argument is based on the following observations:

1. Most von Neumann processors are likely to "idle" during long memory references, and such references are unavoidable in parallel machines.

2. Waits for synchronization events often require task switching, which is expensive on von Neumann machines. Therefore, only certain types of parallelism can be exploited efficiently.

We believe the effect of these issues on performance to be fundamental, and to a large degree, orthogonal to the effect of circuit technology. We will argue that by designing the processor properly, *the detrimental effect of memory latency on performance can be reduced provided there is parallelism in the program*. However, techniques for reducing the effect of latency tend to increase the synchronization cost.

In the rest of this section, we articulate our assumptions regarding general purpose parallel computers. We then discuss the often neglected issue of quantifying the amount of parallelism in programs. Section 2. on page 66 develops a framework for defining the issues of latency and synchronization. Section 3. on page 69 examines the methods to reduce the effect of memory latency in von Neumann computers and discusses their limitations. Section 4. on page 75 similarly examines synchronization methods and their cost. In Section 5. on page 79, we discuss multi-threaded computers like HEP and the MIT Tagged-Token Dataflow machine, and show how these machines can tolerate latency and synchronization costs provided there is sufficient parallelism in programs. The last section summarizes our conclusions.

1.1 Scalable Multiprocessors

We are primarily interested in *general purpose parallel computers*, *i.e.* computers that can exploit parallelism, when present, in any program. Further, we want multiprocessors to be *scalable* in such a manner that adding hardware resources results in higher performance without requiring changes in application programs. The focus of the paper is not on arbitrarily large machines, but machines which range in size from ten to a thousand processors. We expect the processors to be at least as powerful as the current microprocessors and possibly as powerful as the CPU's of the current supercomputers. In particular, the context of the discussion is not machines with millions of one bit ALU's, dozens of which may fit on one chip. The design of such machines will certainly involve fundamental issues in addition to those presented here. Most parallel machines that are available today or likely to be available in the next few years fall within the scope of this paper (*e.g.* BBN Butterfly [36], ALICE [13] and now FLAGSHIP, the Cosmic Cube [38] and Intel's iPSC, IBM's RP3 [33], Alliant and CEDAR [26], and GRIP [11]).

If the programming model of a parallel machine reflects the machine configuration, e.g. number of processors and interconnection topology, the machine is not scalable in a practical sense. Changing the machine configuration should not require changes in application programs or system software; updating tables in the resource management system to reflect the new configuration should be sufficient. However, few multiprocessor designs have taken this stance with regard to scaling. In fact, it is not uncommon to find that source code (and in some cases, algo-

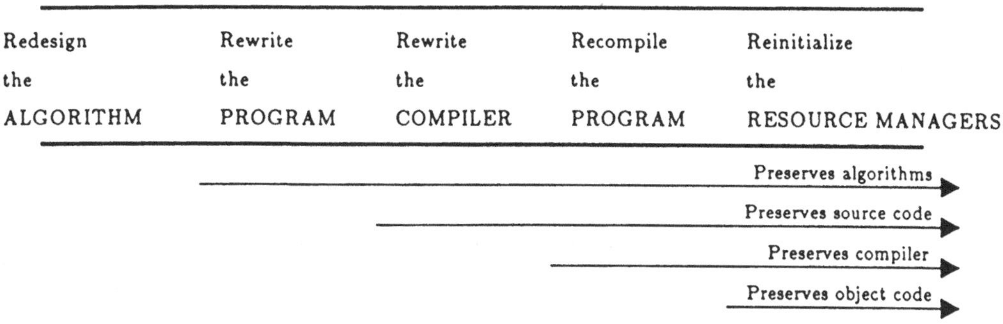

Redesign the ALGORITHM	Rewrite the PROGRAM	Rewrite the COMPILER	Recompile the PROGRAM	Reinitialize the RESOURCE MANAGERS
				Preserves algorithms →
			Preserves source code →	
		Preserves compiler →		
	Preserves object code →			

Figure 1.   The Effect of Scaling on Software

rithms) must be modified in order to run on an altered machine configuration. Figure 1. depicts the range of effects of scaling on the software. Obviously, we consider architectures that support the scenario at the right hand end of the scale to be far more desirable than those at the left. It should be noted that if a parallel machine is not scalable, then it will probably not be fault-tolerant; one failed processor would make the whole machine unusable. It is easy to design hardware in which failed components, e.g. processors, may be masked out. However, if the application code must be rewritten, our guess is that most users would wait for the original machine configuration to be restored.

1.2 Quantifying Parallelism in Programs

Ideally, a parallel machine should speed up the execution of a program in proportion to the number of processors in the machine. Suppose $t(n)$ is the time to execute a program on an n-processor machine. The speed-up as a function of n may be defined as follows:[1]

$$speed - up(n) = \frac{t(1)}{t(n)}$$

Speed-up is clearly dependent upon the program or programs chosen for the measurement. Naturally, if a program does not have "sufficient" parallelism, no parallel machine can be expected to demonstrate dramatic speedup. Thus, in order to evaluate a parallel machine properly, we need to characterize the inherent or potential parallelism of a program. This presents a difficult problem because the amount of parallelism in the source program that is exposed to the architecture may depend upon the quality of the compiler or programmer annotations. Furthermore, there is no reason to assume that the source program cannot be changed. Undoubtedly, different algorithms for a problem have different amounts of parallelism, and the parallelism of an algorithm can be obscured in coding. The

---

[1]   Of course, we are assuming that it is possible to run a program on any number of processors of a machine. In reality often this is not the case.

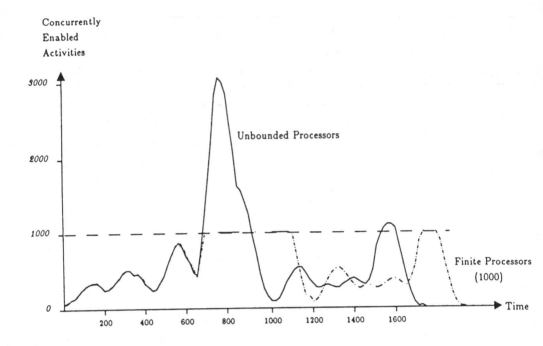

Figure 2. Parallelism Profile of SIMPLE on a 20 x 20 Array

problem is compounded by the fact that most programming languages do not have enough expressive power to show all the possible parallelism of an algorithm in a program. In spite of all these difficulties, we think it is possible to make some useful estimates of the potential parallelism of an algorithm.

It is possible for us to code algorithms in Id [30], a high-level dataflow language, and compile Id programs into dataflow graphs, where the nodes of the graph represent simple operations such as fixed and floating point arithmetic, logicals, equality tests, and memory loads and stores, and where the edges represent only the *essential* data dependencies between the operations. A graph thus generated can be executed on an interpreter (known as GITA) to produce results and the *parallelism profile, pp(t)*, i.e., the number of concurrently executable operators as a function of time on an idealized machine. The idealized machine has unbounded processors and memories, and instantaneous communication. It is further assumed that all operators (instructions) take unit time, and operators are executed as soon as possible. The parallelism profile of a program gives a good estimate of its "inherent parallelism" because it is drawn assuming *the execution of two operators is sequentialized if and only if there is a data dependency between them*. Figure 2 shows the parallelism profile of the SIMPLE code for a representative set of input data. SIMPLE [12], a hydrodynamics and heat flow code kernel, has been extensively studied both analytically [1] and by experimentation.

The solid curve in Figure 2 represents a single outer-loop iteration of SIMPLE on a 20 x 20 mesh, while a typical simulation run performs 100,000 iterations on 100 x 100 mesh. Since there is no significant parallelism between the outer-loop iterations of SIMPLE, the parallelism profile for N iterations can be obtained by repeating the profile in the figure N times. Approximately 75% of the instructions executed involve the usual arithmetic, logical and memory operators; the rest are

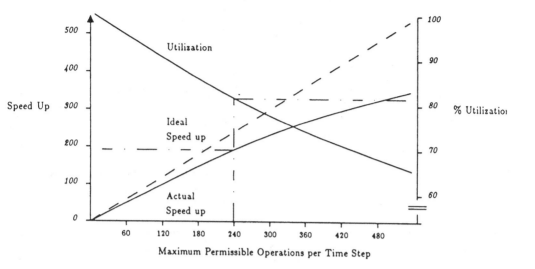

Figure 3. Speed Up and Utilization for 20 x 20 SIMPLE

miscellaneous overhead operators, some of them peculiar to dataflow. One can easily deduce the parallelism profile of any set of operators from the raw data that was used to generate the profile in the figure; however, classifying operators as overhead is not easy in all cases.

The reader may visualize the execution on n processors by drawing a horizontal line at n on the parallelism profile and then "pushing" all the instructions which are above the line to the right and below the line. The dashed curve in Figure 2 shows this for SIMPLE on 1000 processors and was generated by our dataflow graph interpreter by executing the program again with the constraint that no more than n operations were to be performed at any step. However, a good estimate for $t(n)$ can be made, very inexpensively, from the ideal parallelism profile as follows. For any τ, if $pp(\tau) \leq n$, we perform all $pp(\tau)$ operations in time step τ. However, if $pp(\tau) > n$, then we assume it will take the e least integer greater than $pp(\tau)/n$ steps to perform $pp(\tau)$ operations. Hence,

$$t(n) = \sum_{\tau-1}^{T_{max}} \left\lceil \frac{pp(\tau)}{n} \right\rceil$$

where T_{max} is the number of steps in the ideal parallelism profile. Our estimate of $t(n)$ is conservative because the data dependencies in the program may permit the execution of some instructions from $pp(\tau+1)$ in the last time step in which instructions from $pp(\tau)$ are executed.

In our dataflow graphs the number of instructions executed does not change when the program is executed on a different number of processors. Hence, $t(1)$ is simply the area under the parallelism profile. We can now plot $speed-up(n) = t(1)/t(n)$ and $utilization(n) = t(1)/n \times t(n)$, for SIMPLE as shown in Figure 3. For example, in the case of 240 processors, $speed-up$ is 195, and $utilization$ is 81%. One way to understand *utilization(n)* is that a program has *n* parallel operations for only *utilization(n)* fraction of its total *t(n)* duration.

It can be argued that this problem does not have enough parallelism to keep, say, 1000 processors fully utilized. On the other hand, if we cannot keep 10 processors fully utilized, we cannot blame the lack of parallelism in the program. Generally, under-utilization of the machine in the presence of massive parallelism stems from aspects of the internal architecture of the processors which preclude exploitation of certain types of parallelism. Machines are seldom designed to exploit inner-loop, outer-loop, as well as instruction-level parallelism simultaneously.

It is noteworthy that the potential parallelism varies tremendously during execution, a behavior which in our experience is typical of even the most highly parallel programs. We believe that any large program that runs for a long time must have sufficient parallelism to keep hundreds of processors utilized; several applications that we have studied support this belief. However, a parallel machine has to be fairly general purpose and programmable for the user to be able to express even the class of partial differential equation-based simulation programs represented by SIMPLE.

2. Latency and Synchronization

We now discuss the issues of latency and synchronization. We believe latency is most strongly a function of the physical decomposition of a multiprocessor, while synchronization is most strongly a function of how programs are logically decomposed.

2.1 Latency: The First Fundamental Issue

Any multiprocessor organization can be thought of as an interconnection of the following three types of modules (see Figure 4):

1. **Processing elements (PE):** Modules which perform arithmetic and logical operations on data. Each processing element has a single *communication port* through which all data values are received. Processing elements interact with other processing elements by sending messages, issuing interrupts or sending and receiving *synchronizing signals* through shared memory. PE's interact with memory elements by issuing LOAD and STORE instructions modified as necessary with atomicity constraints. Processing elements are characterized by the rate at which they can process instructions. As mentioned, we assume the instructions are simple, *e.g.* fixed and floating point scalar arithmetic. More complex instructions can be counted as multiple instructions for measuring instruction rate.

2. **Memory elements (M):** Modules which store data. Each memory element has a single communication port. Memory elements respond to requests issued by the processing elements by returning data through the communication

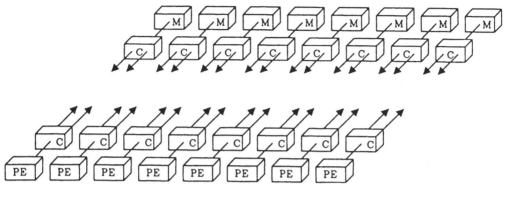

Figure 4.   Structural Model of a Multiprocessor

port, and are characterized by their total capacity and the rate at which they respond to these requests[2]).

3. **Communication elements (C):** Modules which transport data. Each nontrivial communication element has at least three communication ports. Communication elements neither originate nor receive synchronizing signals, instructions, or data; rather, they retransmit such information when received on one of the communication ports to one or more of the other communication ports. Communication elements are characterized by the rate of transmission, the time taken per transmission, and the constraints imposed by one transmission on others, *e.g.* blocking. The maximum amount of data that may be conveyed on a *communication port* per unit time is fixed.

Latency is the time which elapses between making a request and receiving the associated response. The above model implies that *a PE in a multiprocessor system faces larger latency in memory references than in a uniprocessor system* because of the transit time in the communication network between PE's and the memories. The actual interconnection of modules may differ greatly from machine to machine. For example, in the BBN Butterfly machine all memory elements are at an equal distance from all processors, while in IBM's RP3, each processor is closely coupled with a memory element. However, we assume that the average latency in a well designed *n*-PE machine should be $O(log(n))$. In a von Neumann processor, memory latency determines the time to execute memory reference instructions. Usually, the average memory latency also determines the maximum instruction processing speed. When latency cannot be hidden via overlapped operations, a tangible performance penalty is incurred. We call the cost associated with latency as the total *induced processor idle time* attributable to the latency.

---

[2]   In many traditional designs, the "memory" subsystem can be simply modeled by one of these M elements. Interleaved memory subsystems are modeled as a collection of M's and C's. Memory subsystems which incorporate processing capability can be modeled with PE's, M's, and C's. Section 4.3 on page 77 describes one such case.

2.2 Synchronization: The Second Fundamental Issue

We will call the basic units of computation into which programs are decomposed for parallel execution *computational tasks* or simply *tasks*. A general model of parallel programming must assume that tasks are created dynamically during a computation and die after having produced and consumed data. Situations in parallel programming which require task synchronization include the following basic operations:

1. *Producer-Consumer:* A task produces a data structure that is read by another task. If producer and consumer tasks are executed in parallel, synchronization is needed to avoid the *read-before-write* race.

2. *Forks* and *Joins*: The *join* operation forces a synchronization event indicating that two tasks which had been started earlier by some *forking* operation have in fact completed.

3. *Mutual Exclusion:* Non-deterministic events which must be processed one at a time, *e.g.* serialization in the use of a resource.

The minimal support for synchronization can be provided by including instructions, such as atomic TEST-AND-SET, that operate on variables shared by synchronizing tasks[3]). However, to clarify the true cost of such instructions, we will use the *Operational Model* presented in Figure 5. Tasks in the operational model have resources, such as registers and memory, associated with them and constitute the smallest unit of independently schedulable work on the machine. A task is in one of the three states: *ready-to-execute*, *executing* or *suspended*. Tasks ready for execution may be queued locally or globally. When selected, a task occupies a processor until either it completes or is suspended waiting for a synchronization signal. A task changes from *suspended* to *ready-to-execute* when another task causes the relevant synchronization event. Generally, a suspended task must be set aside to avoid deadlocks[4]). The cost associated with such a synchronization is *the fixed time to execute the synchronization instruction plus the time taken to switch to another task*. The cost of task switching can be high because it usually involves saving the processor state, that is, the *context* associated with the task.

There are several subtle issues in accounting for synchronization costs. An event to enable or dispatch a task needs a *name,* such as that of a register or a memory location, and thus, synchronization cost should also include the instructions that generate, match and reuse identifiers which name synchronization events. It may not be easy to identify the instructions executed for this purpose. Nevertheless, such instructions represent overhead because they would not be present if the program were written to execute on a single sequential processor. The hardware design usually dictates the number of names available for synchronization as well as the cost of their use.

The other subtle issue has to do with the accounting for *intra-task synchronization*. As we shall see in Section 3. on page 69, most high performance computers overlap the execution of instructions belonging to one task. The techniques used for

---

[3]   While not strictly necessary, atomic operations such as TEST-AND-SET are certainly a convenient base upon which to build synchronization operations. See Section 4.3 on page 77
[4]   Consider the case of a single processor system which must execute *n* cooperating tasks.

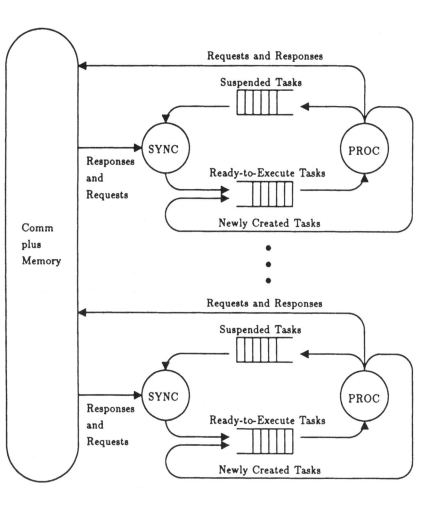

Figure 5.   Operational Model of a Multiprocessor

synchronization of instructions in such a situation (*e.g.* instruction dispatch and suspension) are often quite different from techniques for inter-task synchronization. It is usually safer and cheaper not to put aside the instruction waiting for a synchronization event, but rather to idle (or, equivalently, to execute NO-OP instructions while waiting). This is usually done under the assumption that the idle time will be on the order of a few instruction cycles. We define the synchronization cost in such situations to be the *induced processor idle time* attributable to waiting for the synchronization event.

3. Processor Architectures to Tolerate Latency

In this section, we describe those changes in von Neumann architectures that have directly reduced the effect of memory latency on performance. Increasing the processor state and instruction pipelining are the two most effective techniques for reducing the latency cost. Using Cray-1 (perhaps the best pipelined machine design to date), we will illustrate that it is difficult to keep more than 4 or 5 instructions in the pipeline of a von Neumann processor. It will be shown that every change in the processor architecture which has permitted overlapped execution of instructions has necessitated introduction of a cheap synchronization mechanism. Often these synchronization mechanisms are hidden from the user and not used for inter-task synchronization. This discussion will further illustrate that reducing latency frequently increases synchronization costs.

Before describing these evolutionary changes to hide latency, we should point out that the memory system in a multiprocessor setting creates more problems than just increased latency. Let us assume that all memory modules in a multiprocessor form one global address space and that any processor can read any word in the global address space. This immediately brings up the following problems:

- The time to fetch an operand may not be constant because some memories may be "closer" than others in the physical organization of the machine.

- No useful bound on the worst case time to fetch an operand may be possible at machine design time because of the scalability assumption. This is at odds with RISC designs which treat memory access time as bounded and fixed.

- If a processor were to issue several (pipelined) memory requests to different remote memory modules, the responses could arrive out of order.

All of these issues are discussed and illustrated in the following sections. A general solution for accepting memory responses out of order requires a synchronization mechanism to match responses with the destination registers (*names* in the task's context) and the instructions waiting on that value. The ill-fated Denelcor HEP [25] is one of the very few architectures which has provided such mechanisms in the von Neumann framework. However, the architecture of the HEP is sufficiently different from von Neumann architectures as to warrant a separate discussion (see Section 5. on page 79).

3.1 Increasing the Processor State

Figure 6 depicts the modern-day view of the von Neumann computer [9] (*sans* I/O). In the earliest computers, such as EDSAC, the *processor state* consisted solely of an accumulator, a quotient register, and a program counter. Memories were relatively slow compared to the processors, and thus, the time to fetch an instruction and its operands completely dominated the instruction cycle time. Speeding up the Arithmetic Logic Unit was of little use unless the memory access time could also be reduced.

The appearance of multiple "accumulators" reduced the number of operand fetches and stores, and index registers dramatically reduced the number of instructions executed by essentially eliminating the need for self-modifying code.

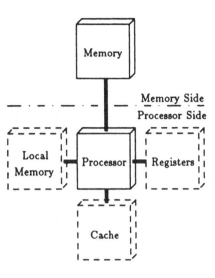

Figure 6.   The von Neumann Processor *(from Gajski and Peir [20])*

Since the memory traffic was drastically lower, programs executed much faster than before. However, the enlarged processor state did not reduce the time lost during memory references and, consequently, did not contribute to an overall reduction in cycle time; the basic cycle time improved only with improvements in circuit speeds.

3.2 Instruction Prefetching

The time taken by instruction fetch (and perhaps part of instruction decoding time) can be totally hidden if prefetching is done during the execution phase of the previous instruction. If instructions and data are kept in separate memories, it is possible to overlap instruction prefetching and operand fetching also. (The IBM STRETCH [7] and Univac LARC [16] represent two of the earliest attempts at implementing this idea.) Prefetching can reduce the cycle time of the machine by twenty to thirty percent depending upon the amount of time taken by the first two steps of the instruction cycle with respect to the complete cycle. However, the effective throughput of the machine cannot increase proportionately because overlapped execution is not possible with *all* instructions.

Instruction prefetching works well when the execution of instruction n does not have any effect on either the choice of instructions to fetch (as is the case in BRANCH) or the content of the fetched instruction (self-modifying code) for instructions $n + 1, n + 2, \dots, n + k$. The latter case is usually handled by simply outlawing it. However, effective overlapped execution in the presence of BRANCH instructions has remained a problem. Techniques such as prefetching both BRANCH targets have shown little performance/cost benefits. Lately, the concept of *delayed* BRANCH instructions from microprogramming has been incorporated, with success, in LOAD/STORE architectures (see 3.4 on page 73). The idea is to delay the effect of a BRANCH by one instruction. Thus, the instruction at $n + 1$ following a BRANCH instruction at n is always executed regardless of which way the BRANCH at n goes. One can always follow a BRANCH instruction with a NO-

OP instruction to get the old effect. However, experience has shown that seventy percent of the time a useful instruction can be put in that position.

3.3 Instruction Buffers, Operand Caches and Pipelined Execution

The time to fetch instructions can be further reduced by providing a fast instruction buffer. In machines such as the CDC 6600 [40] and the Cray-1 [37], the instruction buffer is automatically loaded with n instructions in the neighborhood of the referenced instruction (relying on spatial locality in code references), whenever the referenced instruction is found to be missing. To take advantage of instruction buffers, it is also necessary to speed up the operand fetch and execute phases. This is usually done by providing *operand* caches or buffers, and overlapping the operand fetch and execution phases[5] Of course, balancing the pipeline under these conditions may require further pipelining of the ALU. If successful, these techniques can reduce the machine cycle time to one-fourth or one-fifth the cycle time of an unpipelined machine. However, overlapped execution of four to five instructions in the von Neumann framework presents some serious conceptual difficulties, as discussed next.

Designing a well-balanced pipeline requires that the time taken by various pipeline stages be more or less equal, and that the "things", *i.e.* instructions, entering the pipe be independent of each other. Obviously, instructions of a program cannot be totally independent except in some special trivial cases. Instructions in a pipe are usually related in one of two ways: Instruction n produces data needed by instruction $n+k$, or only the complete execution of instruction n determines the next instruction to be executed (the aforementioned BRANCH problem).

Limitations on hardware resources can also cause instructions to interfere with one another. Consider the case when both instructions n and $n + 1$ require an adder, but there is only one of these in the machine. Obviously, one of the instructions must be deferred until the other is complete. A pipelined machine must be temporarily able to prevent a new instruction from entering the pipeline when there possibility of interference with the instructions already in the pipe. Detecting and quickly resolving these *hazards* is very difficult with ordinary instruction sets, *e.g.*, IBM 370, VAX 11 or Motorola 68000, due to their complexity.

A major complication in pipelining complex instructions is the variable amount of time taken in each stage of instruction processing (refer to Figure 7). Operand fetch in the VAX is one such example: determining the addressing mode for each operand requires a fair amount of decoding, and actual fetching can involve 0 to 2 memory references per operand. Considering all possible addressing mode combinations, an instruction may involve 0 to 6 memory references in addition to the instruction fetch itself! A pipeline design that can effectively tolerate such variations is close to impossible.

---

[5]   As we will show in Section 4.4 on page 78, caches in a multiprocessor setting create special problems.

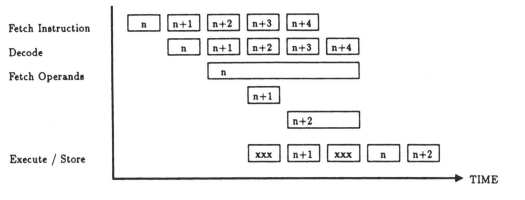

Figure 7.   Variable Operand Fetch Time

3.4 Load/Store Architectures

Seymour Cray, in the sixties, pioneered instruction sets (CDC 6600, Cray-1) which separate instructions into two disjoint classes. In one class are instructions which move data *unchanged* between memory and high speed registers. In the other class are instructions which operate on data in the registers. Instructions of the second class *cannot* access the memory. This rigid distinction simplifies instruction scheduling. For each instruction, it is trivial to see if a memory reference will be necessary or not. Moreover, the memory system and the ALU may be viewed as parallel, noninteracting pipelines. An instruction dispatches exactly one unit of work to either one pipe or the other, but never both.

Such architectures have come to be known as LOAD/STORE architectures, and include the machines built by Reduced Instruction Set Computer (RISC) enthusiasts (the IBM 801 [34], Berkeley's RISC [32], and Stanford MIPS [22] are prime examples). LOAD/STORE architectures use the time between instruction decoding and instruction dispatching for hazard detection and resolution (see Figure 8). The design of the instruction pipeline is based on the principle that if an instruction gets past some fixed pipe stage, it should be able to run to completion without incurring any previously unanticipated hazards.

LOAD/STORE architectures are much better at tolerating latencies in memory accesses than other von Neumann architectures. In order to explain this point, we will first discuss a simplified model which detects and avoids hazards in a LOAD/STORE architecture similar to the Cray-1. Assume there is a bit associated with every register to indicate that the contents of the register are undergoing a change. The bit corresponding to register R is set the moment we dispatch an instruction that wants to update R. Following this, instructions are allowed to enter the pipeline only if they don't need to reference or modify register R or other registers reserved in a similar way. Whenever a value is stored in R, the reservation on R is removed, and if an instruction is waiting on R, it is allowed to proceed. This simple scheme works only if we assume that registers whose values are needed by an instruction are read before the next instruction is dispatched, and that the ALU or the multiple functional units within the ALU are pipelined to accept inputs

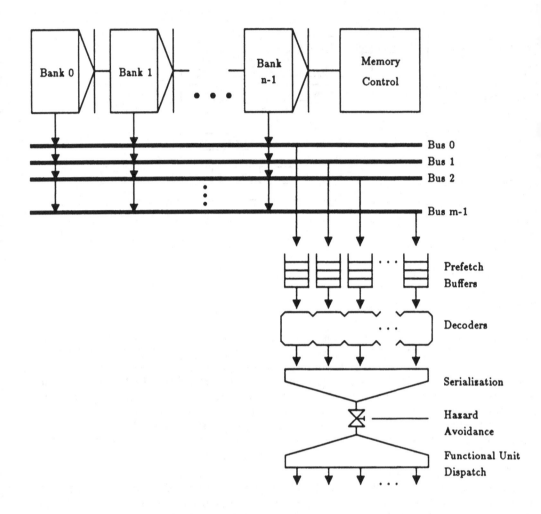

Figure 8.   Hazard Avoidance at the Instruction Decode Stage

as fast as the decode stage can supply them[6]). The dispatching of an instruction can also be held up because it may require a bus for storing results in a clock cycle when the bus is needed by another instruction in the pipeline. Whenever BRANCH instructions are encountered, the pipeline is effectively held up until the branch target has been decided.

Notice what will happen when an instruction to load the contents of some memory location M into some register R is executed. Suppose that it takes k cycles to fetch something from the memory. It will be possible to execute several instructions during these k cycles as long as none of them refer to register R. In fact, this situation is hardly different from the one in which R is to be loaded from some func-

---

tional unit that, like the Floating Point multiplier, takes several cycles to produce the result. These gaps in the pipeline can be further reduced if the compiler reorders instructions such that instructions consuming a datum are put as far as possible from instructions producing that datum. Thus, we notice that machines designed for high pipelining of instructions can hide large memory latencies provided there is local parallelism among instructions[7].

From another point of view, latency cost has been reduced by introducing a cheap synchronization mechanism: reservation bits on processor registers. However, the number of *names* available for synchronization, *i.e.* the size of the task's processor-bound context, is precisely the number of registers, and this restricts the amount of exploitable parallelism and tolerable latency. In order to understand this issue better, consider the case when the compiler decides to use register R to hold two different values at two different instructions say, i_n and i_m. This will require i_n and i_m to be executed sequentially while no such order may have been implied by the source code. *Shadow registers* have been suggested to deal with this class of problems. In fact, shadow registers are an engineering approach to solving a non-engineering problem. The real issue is *naming*. The reason that addition of explicit and implicit registers improves the situation derives from the addition of (explicit and implicit) *names* for synchronization and, hence, a greater opportunity for tolerating latency.

Some LOAD/STORE architectures have eliminated the need for reservation bits on registers by making the compiler responsible for scheduling instructions, such that the result is guaranteed to be available. The compiler can perform hazard resolution only if the time for each operation *e.g.* ADD, LOAD, is known; it inserts NO-OP instructions wherever necessary. Because the instruction execution times are an intimate part of the object code, *any* change to the machine's structure (scaling, redesign) will at the very least require changes to the compiler and regeneration of the code. This is obviously contrary to our notion of generality, and hinders the portability of software from one generation of machine to the next.

Current LOAD/STORE architectures assume that memory references either take a fixed amount of time (one cycle in most RISC machines) or that they take a variable but predictable amount of time (as in the Cray-1). In RISC machines, this time is derived on the basis of a cache hit. If the operand is found to be missing from the cache, the pipeline stops. Equivalently, one can think of this as a situation where a clock cycle is *stretched* to the time required. This solution works because, in most of these machines, there can be either one or a very small number of memory references in progress at any given time. For example, in the Cray-1, no more than four independent addresses can be generated during a memory cycle. If the generated address causes a bank conflict, the pipeline is stopped. However, any conflict is resolved in at most three cycles.

LOAD/STORE architectures, because of their simpler instructions, often execute 15% to 50% more instructions than machines with more complex instructions [34]. This increase may be regarded as synchronization cost. However, this is easily compensated by improvements in clock speed made possible by simpler control mechanisms.

---

[7] The ability to reorder two instructions usually means that these instructions can be executed in parallel.

4. Synchronization Methods for Multiprocessing

4.1 Global Scheduling on Synchronous machines

For a totally synchronous multiprocessor it is possible to envision a master plan which specifies operations for every cycle on every processor. An analogy can be made between programming such a multiprocessor and coding a horizontally microprogrammed machine. Recent advances in compiling [18] have made such code generation feasible and encouraged researchers to propose and build several different synchronous multiprocessors. Cydrome and Multiflow computers, which are based on proposals in [35] and [19], respectively, are examples of such machines. These machines are generally referred to as *very long instruction word*, or VLIW, machines, because each instruction actually contains multiple smaller instructions (one per functional unit or processing element). The strategy is based on maximizing the use of resources and resolving potential run-time conflicts in the use of resources at compile time. Memory references and control transfers are "anticipated" as in RISC architectures, but here, multiple concurrent threads of computation are being scheduled instead of only one. Given the possibility of decoding and initiating many instructions in parallel, such architectures are highly appealing when one realizes that the fastest machines available now still essentially decode and dispatch instructions one at a time.

We believe that this technique is effective in its currently realized context, *i.e.* Fortran-based computations on a small number (4 to 8) of processors. Compiling for parallelism beyond this level, however, becomes intractable. It is unclear how problems which rely on dynamic storage allocation or require nondeterministic and real-time constraints will play out on such architectures.

4.2 Interrupts and Low-level Context Switching

Almost all von Neumann machines are capable of accepting and handling interrupts. Not surprisingly, multiprocessors based on such machines permit the use of inter-processor interrupts as a means for signalling events. However, interrupts are rather expensive because, in general, the processor state needs to be saved. The state-saving may be forced by the hardware as a direct consequence of allowing the interrupt to occur, or it may occur explicitly, *i.e.* under the control of the programmer, via a single very complex instruction or a suite of less complex ones. Independent of *how* the state-saving happens, the important thing to note is that each interrupt will generate a significant amount of traffic across the processor - memory interface.

In the previous discussion, we concluded that larger processor state is good because it provided a means for reducing memory latency cost. In trying to solve the problem of low cost synchronization, we have now come across an interaction which, we believe, is more than just coincidental. Specifically, in very fast von Neumann processors, the "obvious" synchronization mechanism (interrupts) will only work well in the trivial case of infrequent synchronization events or when the amount of processor state which must be saved is *very small*. Said another way, reducing the cost of synchronization by making interrupts cheap would generally entail increasing the cost of memory latency.

Uniprocessors such as the Xerox Alto [41], the Xerox Dorado [27], and the Symbolics 3600 family [29] have used a technique which may be called *microcode-level context switching* to allow sharing of the CPU resource by the I/O device adapters. This is accomplished by duplicating programmer-visible registers, in other words, the processor state. Thus, in one microinstruction the processor can be switched to a new task without causing any memory references to save the processor state[8]) This dramatically reduces the cost of processing certain types of events that cause frequent interrupts. As far as we know, nobody has adapted the idea of keeping multiple contexts in a multiprocessor setting (with the possible exception of the HEP, to be discussed in Section 5. on page 79) although it should reduce synchronization cost over processors which can hold only a single context. It may be worth thinking about adopting this scheme to reduce the latency cost of a nonlocal memory references as well.

The limitations of this approach are obvious. High performance processors may have a small programmer-visible state (number of registers) but a much larger implicit state (caches). Low-level task switching does not necessarily take care of the overhead of flushing caches[9]). Further, one can only have a small number of independent contexts without completely overshadowing the cost of ALU hardware.

4.3 Semaphores and the Ultracomputer

Next to interrupts, the most commonly supported feature for synchronization is an *atomic operation* to test and set the value of a memory location. A processor can signal another processor by writing into a location which the other processor keeps reading to sense a change. Even though, theoretically, it is possible to perform such synchronization with ordinary read and write memory operations, the task is much simpler with an atomic TEST-AND-SET instruction. TEST-AND-SET is powerful enough to implement all types of synchronization paradigms mentioned earlier. However, the synchronization cost of using such an instruction can be very high. Essentially, the processor that executes it goes into a *busy-wait* cycle. Not only does the processor get blocked, it generates extra memory references at every instruction cycle until the TEST-AND-SET instruction is executed successfully. Implementations of TEST-AND-SET that permit non-busy waiting imply context switching in the processor and thus are not necessarily cheap either.

It is possible to improve upon the TEST-AND-SET instruction in a multiprocessor setting, as suggested by the NYU Ultracomputer group [17]. Their technique can be illustrated by the atomic FETCH-AND-⟨OP⟩ instruction (an evolution of the REPLACE-ADD instruction). The instruction requires an address and a value, and works as follows: suppose two processors, i and j, simultaneously execute FETCH-AND-ADD instructions with arguments (A,v_i) and (A,v_j) respectively. After one instruction cycle, the contents of A will become $(A) + v_i + v_j$. Processors i and j will receive, respectively, either (A) and $(A) + v_i$, or $(A) + v_j$ and (A) as results. Indeterminacy is a direct consequence of the race to update memory cell A.

---

[8]) The Berkeley RISC idea of providing "register windows" to speed up procedure calls is very similar to multiple contexts.

[9]) However, solutions such as multicontext caches and multicontext address translation buffers have been used to advantage in reducing this task switching overhead, (c.f., the *STO* stack mechanism in the IBM 370/168).

An architect must choose between a wide variety of implementations for FETCH-AND-⟨OP⟩. One possibility is that the processor may interpret the instruction with a series of more primitive instructions. While possible, such a solution does not find much favor because it will cause considerable memory traffic. A second scheme implements FETCH-AND-⟨OP⟩ in the memory controller (this is the alternative chosen by the CEDAR project [28]). This typically results in a significant reduction of network traffic because atomicity of memory transactions from the memory's controller happens by default. The scheme suggested by the NYU Ultracomputer group implements the instruction *in the switching nodes of the network*.

This implementation calls for a *combining* packet communication network which connects n processors to an n-port memory. If two packets collide, say FETCH-AND-ADD(A,v_i) and FETCH-AND-ADD(A,v_j), the switch extracts the values v_i and v_j, forms a new packet (FETCH-AND-ADD($A,v_i + v_j$)) , forwards it to the memory, and stores the value of v_i temporarily. When the memory returns the old value of location A, the switch returns two values $((A)$ and $(A) + v_i)$. The main improvement is that some synchronization situations which would have taken $O(n)$ time can be done in $O(\log n)$ time. It should be noted, however, that one memory reference may involve as many as $\log_2 n$ additions, and implies substantial hardware complexity. Further, the issue of processor idle time due to latency has not been addressed at all. In the worst case, the complexity of hardware may actually increase the latency of going through the switch and thus completely overshadow the advantage of "combining" over other simpler implementations.

The simulation results reported by NYU [17] show quasi-linear speedup on the Ultracomputer (a shared memory machine with ordinary von Neumann processors, employing FETCH-AND-ADD synchronization) for a large variety of scientific applications. We are not sure how to interpret these results without knowing many more details of their simulation model. Two possible interpretations are the following:

1. Parallel branches of a computation hardly share any data, thus, the costly *mutual exclusion* synchronization is rarely needed in real applications.

2. The synchronization cost of using shared data can be acceptably brought down by judicious use of cachable/non cachable annotations in the source program.

The second point may become clearer after reading the next section.

4.4 Cache Coherence Mechanisms

While highly successful for reducing memory latency in uniprocessors, caches in a multiprocessor setting introduce a serious synchronization problem called *cache coherence*. Censier and Feautrier [10] define the problem as follows: "*A memory scheme is coherent if the value returned on a LOAD instruction is always the value given by the latest STORE instruction with the same address.*". It is easy to see that this may be difficult to achieve in multiprocessing.

Suppose we have a two-processor system tightly coupled through a single main memory. Each processor has its own cache to which it has exclusive access. Suppose further that two tasks are running, one on each processor, and we know that the tasks are designed to communicate through one or more shared memory cells.

In the absence of caches, this scheme can be made to work. However, if it happens that the shared address is present in both caches, the individual processors can read and write the address and *never* see any changes caused by the other processor. Using a store-through design instead of a store-in design does not solve the problem either. What is logically required is a mechanism which, upon the occurrence of a STORE to location x, invalidates copies of location x in caches of other processors, and guarantees that subsequent LOAD will get the most recent (cached) value. This can incur significant overhead in terms of decreased memory bandwidth.

All solutions to the cache coherence problem center around reducing the cost of detecting rather than avoiding the possibility of cache incoherence. Generally, *state* information indicating whether the cached data is private or shared, read-only or read-write, etc., is associated with each cache entry. However, this state somehow has to be updated after each memory reference. Implementations of this idea are generally intractable except possibly in the domain of bus-oriented multiprocessors. The so-called *snoopy bus* solution uses the broadcasting capability of buses and purges entry x from all caches when a processor attempts a STORE to x. In such a system, at most one STORE operation can go on at a time in the whole system and, therefore, system performance is going to be a strong function of the snoopy bus' ability to handle the coherence-maintaining traffic.

It is possible to improve upon the above solution if some additional state information is kept with each cache entry. Suppose entries are marked "shared" or "non-shared". A processor can freely read shared entries, but an attempt to STORE into a shared entry immediately causes that address to appear on the snoopy bus. That entry is then deleted from all the other caches and is marked "non-shared" in the processor that had attempted the STORE. Similar action takes place when the word to be written is missing from the cache. Of course, the main memory must be updated before purging the private copy from any cache. When the word to be read is missing from the cache, the snoopy bus may have to first reclaim the copy privately held by some other cache before giving it to the requesting cache. The status of such an entry will be marked as shared in both caches. The advantage of keeping shared/non-shared information with every cache entry is that the snoopy bus comes into action only on cache misses and STOREs to shared locations, as opposed to all LOADs and STOREs. Even if these solutions work satisfactorily, bus-oriented multiprocessors are not of much interest to us because of their obvious limitations in scaling.

As far as we can tell, there are no known solutions to cache coherence for non-bussed machines. It would seem reasonable that one needs to make caches partially visible to the programmer by allowing him to mark data (actually addresses) as shared or not shared. In addition, instructions to flush an entry or a block of entries from a cache have to be provided. Cache management on such machines is possible only if the concept of shared data is well integrated in the high-level language or the programming model. Schemes have also been proposed explicitly to interlock a location for writing or to bypass the cache (and flush it if necessary) on a STORE; in either case, the performance goes down rapidly as the machine is scaled. Ironically, in solving the latency problem via multiple caches, we have introduced the synchronization problem of keeping caches coherent.

It is worth noting that, while not obvious, a direct trade-off often exists between decreasing the parallelism and increasing the cachable or non-shared data.

5. Multi-Threaded Architectures

In order to reduce memory latency cost, it is essential that a processor be capable of issuing multiple, overlapped memory requests. The processor must view the memory/communication subsystems as a logical pipeline. As latency increases, keeping the pipeline full implies that more memory references will have to be in the pipeline. We note that memory systems of current von Neumann architectures have very little capability for pipelining, with the exception of array references in vector machines. The reasons behind this limitation are fundamental:

1. von Neumann processors must observe instruction sequencing constraints, and

2. since memory references can get out of order in the pipeline, a large number of identifiers to distinguish memory responses must be provided.

One way to overcome the first deficiency is to interleave many threads of sequential computations (as we saw in the very long instruction word architectures of 4.1 on page 75). The second deficiency can be overcome by providing a large register set with suitable reservation bits. It should be noted that these requirements are somewhat in conflict. The situation is further complicated by the need of tasks to communicate with each other. Support for cheap synchronization calls for the processor to switch tasks quickly and to have a non-empty queue of tasks which are ready to run. One way to achieve this is again by interleaving multiple threads of computation and providing some intelligent scheduling mechanism to avoid busy-waits. Machines supporting multiple threads and fancy scheduling of instructions or processes look less and less like von Neumann machines as the number of threads increases.

In this section, we first discuss the erstwhile Denelcor HEP [25, 39]. The HEP was the first commercially available multi-threaded computer. After that we briefly discuss dataflow machines, which may be regarded as an extreme example of machines with multiple threads; machines in which each instruction constitutes an independent thread and only non-suspended threads are scheduled to be executed.

5.1 The Denelcor HEP: A Step Beyond von Neumann Architectures

The basic structure of the HEP processor is shown in Figure 9. The processor's data path is built as an eight step pipeline. In parallel with the data path is a control loop which circulates process status words (PSW's) of the processes whose threads are to be interleaved for execution. The delay around the control loop varies with the queue size, but is never shorter than eight pipe steps. This minimum value is intentional to allow the PSW at the head of the queue to initiate an instruction but not return again to the head of the queue until the instruction has completed. If at least eight PSW's, representing eight processes, can be kept in the queue, the processor's pipeline will remain full. This scheme is much like traditional pipelining of instructions, but with an important difference. The inter-instruction dependencies are likely to be weaker here because adjacent instructions in the pipe are always from *different processes*.

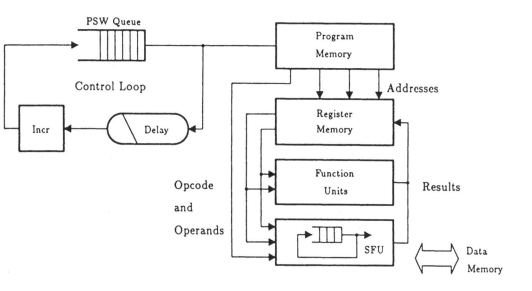

Figure 9.   Latency Toleration and Synchronization in the HEP

There are 2048 registers in each processor; each process has an index offset into the register array. Inter-process, *i.e.*, inter-thread, communication is possible via these registers by overlapping register allocations. The HEP provides FULL/EMPTY/RESERVED bits on each register and FULL/EMPTY bits on each word in the data memory. An instruction encountering EMPTY or RESERVED registers behaves like a NO-OP instruction; the program counter of the process, *i.e.* PSW, which initiated the instruction is not incremented. The process effectively *busy-waits* but without blocking the processor. When a process issues a LOAD or STORE instruction, it is removed from the control loop and is queued separately in the Scheduler Function Unit (SFU) which also issues the memory request. Requests which are not satisfied because of improper FULL/EMPTY status result in recirculation of the PSW within the SFU's loop and also in reissuance of the request. The SFU matches up memory responses with queued PSW's, updates registers as necessary and reinserts the PSW's in the control loop.

Thus, the HEP is capable up to a point of using parallelism in programs to hide memory and communication latency. At the same time it provides efficient, low-level synchronization mechanisms in the form of presence-bits in registers and main memory. However, the HEP approach does not go far enough because there is a limit of *one* outstanding memory request per process, and the cost of synchronization through shared registers can be high because of the loss of processor time due to *busy-waiting*. A serious impediment to the software development on HEP was the limit of 64 PSW's in each processor. Though only 8 PSW's may be required to keep the process pipeline full, a much larger number is needed to name all concurrent tasks of a program.

5.2 Dataflow Architectures

Dataflow architectures [2, 15, 21, 23] represent a radical alternative to von Neumann architectures because they use dataflow graphs as their machine language [4, 14]. Dataflow graphs, as opposed to conventional machine languages, specify only a partial order for the execution of instructions and thus provide opportunities for parallel and pipelined execution at the level of individual instructions. For example, the dataflow graph for the expression $a \cdot b + c \cdot d$ only specifies that both multiplications be executed before the addition; however, the multiplications can be executed in any order or even in parallel. The advantage of this flexibility becomes apparent when we consider that the order in which a, b, c and d will become available may not be known at compile time. For example, computations for operands a and b may take longer than computations for c and d or *vice versa*. Another possibility is that the time to fetch different operands may vary due to scheduling and hardware characteristics of the machine. Dataflow graphs do not force unnecessary sequentialization and dataflow processors schedule instructions according to the availability of the operands.

The instruction execution mechanism of a dataflow processor is fundamentally different from that of a von Neumann processor. We will briefly illustrate this using the MIT Tagged-Token architecture (see Figure 10). Rather than following a *Program Counter* for the next instruction to be executed and then fetching operands for that instruction, a dataflow machine provides a low-level synchronization mechanism in the form of *Waiting-Matching* section which dispatches only those instructions for which data are already available. This mechanism relies on *tagging* each datum with the address of the instruction to which it belongs and the context in which the instruction is being executed. One can think of the instruction address as replacing the program counter, and the context identifier replacing the frame base register in traditional von Neumann architecture. It is the machine's job to match up data with the same tag and then to execute the denoted instruction. In so doing, new data will be produced, with a new tag indicating the successor instruction(s). Thus, each instruction represents a synchronization operation. Note that the number of synchronization names is limited by the size of the tag, which can easily be made much larger than the size of the register array in a von Neumann machine. Note also that the processor pipeline is non-blocking: given that the operands for an instruction are available, the corresponding instruction can be executed without further synchronization.

In addition to the waiting-matching section which is used primarily for dynamic scheduling of instructions, the MIT Tagged-Token machine provides a second synchronization mechanism called *I-Structure Storage*. Each word of I-structure storage has 2 bits associated with it to indicate whether the word is empty, full or has pending read-requests. This greatly facilitates overlapped execution of a producer of a data structure with the consumer of that data structure. There are three instructions at the graph level to manipulate I-structure storage. These are *allocate* - to allocate n empty words of storage, *select* - to fetch the contents of the ith word of an array and *store* - to store a value in a specified word. Generally software concerns dictate that a word be written into only once before it is deallocated. The dataflow processor treats all I-structure operations as *split-phase*. For example, when the *select* instruction is executed, a packet containing the tag of the destination instruction of the select instruction is forwarded to the proper address, possibly in a distant I-structure storage module. The actual memory operation may

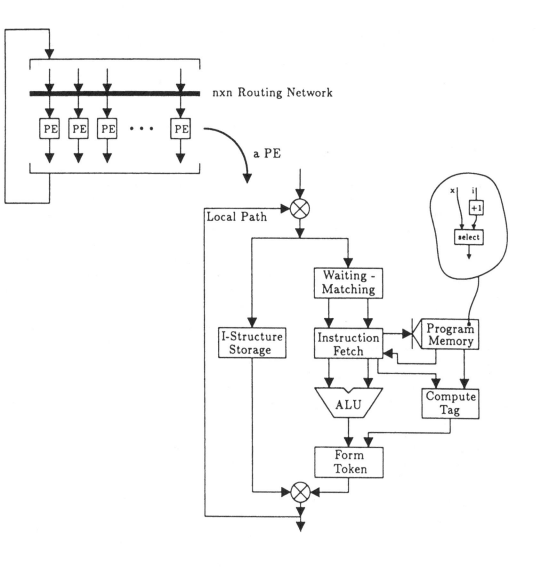

Figure 10. The MIT Tagged-Token Dataflow Machine

require waiting if the data is not present and thus the result may be returned many instruction times later. The key is that the instruction pipeline need not be suspended during this time. Rather, processing of other instructions may continue immediately after *initiation* of the operation. Matching of memory responses with waiting instructions is done via tags in the waiting-matching section.

One advantage of tagging each datum is that data from different contexts can be mixed freely in the instruction execution pipeline. Thus, instruction-level parallelism of dataflow graphs can effectively absorb the communication latency and minimize the losses due to synchronization waits. We hope it is clear from the prior discussion that even the most highly pipelined von Neumann processor cannot match the flexibility of a dataflow processor in this regard. A more complete discussion of dataflow machines is beyond the scope of this paper. An overview of executing programs on the MIT Tagged-Token Dataflow machine can be found in [6]. A deeper understanding of dataflow machines can be gotten from [2]. Additional, albeit slightly dated, details of the machine and the instruction set are given in [3] and [5], respectively.

6. Conclusions

We have presented the loss of performance due to increased latency and waits for synchronization events as the two fundamental issues in the design of parallel machines. These issues are, to a large degree, independent of the technology differences between various parallel machines. Even though we have not presented it as such, these issues are also independent of the high-level programming model used on a multiprocessor. If a multiprocessor is built out of conventional microprocessors, then degradation in performance due to latency and synchronization will show up regardless of whether a shared-memory, message-passing, reduction or dataflow programming model is employed.

Is it possible to modify a von Neumann processor to make it more suitable as a building block for a parallel machine? In our opinion the answer is a qualified "yes". The two most important characteristics of the dataflow processor are split-phase memory operations and the ability to put aside computations (i.e., processes, instructions, or whatever the scheduling quanta are) without blocking the processor. We think synchronization bits in the storage are essential to support the producer-consumer type of parallelism. However, the more concurrently active threads of computation we have, the greater is the requirement for hardware-supported synchronization names. Iannucci [24] and others [8] are actively exploring designs based on these ideas. Only time will tell if it will be fair to classify such processors as von Neumann processors.

The biggest appeal of von Neumann processors is that they are widely available and familiar. There is a tendency to extrapolate these facts into a belief that von Neumann processors are "simple" and efficient. A technically sound case can be made that well designed von Neumann processors are indeed very efficient in executing sequential codes and require less memory bandwidth than dataflow processors. However, the efficiency of sequential threads disappears fast if there are too many interruptions or if idling of the processor due to latency or data-dependent hazards increases. Papadopoulos [31] is investigating dataflow architectures which will improve the efficiency of the MIT Tagged-Token architecture on sequential codes without sacrificing any of its dataflow advantages. We can assure the reader that none of these changes are tantamount to introducing a program counter in the dataflow architecture.

For lack of space we have not discussed the effect of multi-threaded architectures on the compiling and language issues. It is important to realize that compiling into primitive dataflow operators is a much simpler task than compiling into cooperating sequential threads. Since the cost of inter-process communication in a von Neumann setting is much greater than the cost of communication within a process, there is a preferred process or "grain" size on a given architecture. Furthermore, placement of synchronization instructions in a sequential code requires careful planning because an instruction to wait for a synchronization event may experience very different waiting periods in different locations in the program. Thus, even for a given grain size, it is difficult to decompose a program optimally. Dataflow graphs, on the other hand, provide a uniform view of inter- and intra-procedural synchronization and communication, and as noted earlier, only specify a partial order to enforce data dependencies among the instructions of a program. Though it is very difficult to offer a quantitative measure, we believe that an Id Nouveau compiler to generate code for a multi-threaded von Neumann computer will be significantly more complex than the current compiler [41] which generates fine grain dataflow graphs for the MIT Tagged-Token dataflow machine. Thus dataflow computers, in addition to providing solutions to the fundamental hardware issues raised in this paper, also have compiler technology to exploit their full potential.

Acknowledgment

The authors wish to thank David Culler for valuable discussions on much of the subject matter of this paper, particularly Load/Store architectures and the structure of the Cray machines. Members of the Computation Structures Group have developed many tools, without which the analysis of the Simple code would have been impossible. In particular, we would like to thank Ken Traub for the ID Compiler and David Culler and Dinarte Morais for GITA. This paper has benefited from numerous discussions with people both inside and outside MIT. We wish to thank Natalie Tarbet, Ken Traub, David Culler, Vinod Kathail and Rishiyur Nikhil for suggestions to improve this manuscript.

7. References

[1]   Arvind and R.E. Bryant
       Design Considerations for a Partial Equation Machine. Proceedings of Scientific
       Computer Information Exchange Meeting,
       Lawrence Livermore Laboratory, Livermore, CA, September, 1979,
       pp. 94-102.

[2]   Arvind and D.E. Culler
       "Dataflow Architectures".
       Annual Reviews of Computer Science I (1986), 225-253.

[3]   Arvind and D.E. Culler, R.A. Iannucci, V. Kathail, K. Pingali, and R.E. Thomas
       The Tagged Token Dataflow Architecture.
       Internal Report. (including architectural revisions of October, 1983).

[4]   Arvind and K.P. Gostelow
       "The U-Interpreter".
       Computer 15, 2 (February 1982), 42-49.

[5]   Arvind and R.A. Iannucci
       Instruction Set Definition for a Tagged-Token Data FLow Machine.
       Computation Structures Group Memo 212-3, Laboratory for Computer Science,
       MIT, Cambridge, Mass., Cambridge, MA 02139, December, 1981.

[6]   Arvind and R.S. Nikhil
       Executing a Program on the MIT Tagged-Token Data FLow Architecture.
       Proc. PARLE, (Parallel Architectures and Languages Europe), Eindhoven, The
       Netherlands, June, 1987.

[7]   Block, E.
        The Engineering Design of the STRETCH Computer.
        Proceedings of the EJCC, 1959, pp. 48-59.

[8]   Buehrer, R. and K. Ekanadham
        Dataflow Principles in Multi-processor Systems.
        ETH Zurich, and Research Division, Yorktown Heights, IBM Corporation, July,
        1986.

[9]   Burks, A., H.H. Goldstine, and J. von Neumann
        "Preliminary Discussion of the Logical Design of an Electronic Instrument, Part
        2".
        Datamation 8, 10 (October 1962), 36-41,

[10]   Censier, L.M. and P. Feautrier
        "A New Solution to the Coherence Problems in Multicache Systems".
        IEEE Transactions on Computers C-27, 12 (December 1979), 1112-1118.

[11]   Clack, C. and Peyton-Jones, S.L.
        The Four-Stroke Reduction Engine.
        Proceedings of the 1986 ACM Conference on Lisp and Functional Programming,
        Association for Computing Machinery, August, 1986, pp. 220-232.

[12]   Crowley, W.P., C.P. Hendrickson and T.E. Rudy
        The SIMPLE Code.
        Internal Report UCID-17715, Lawrence Livermore Laboratory, Livermore, CA, Feb-
        ruary, 1978.

[13]   Darlington, J. and M Reeve
        ALICE: A Multi-Processor Reduction Machine for the Parallel Evaluation of Appli-
        cative Languages.
        Proceedings of the 1981 Conference on Functional Programming Languages and
        Computer Architecture, Portsmouth, NH, 1981, pp. 65-76.

[14]   Dennis, J.B.
        Lecture Notes in Computer Science. Volume 19: First Version of a Data Flow Pro-
        cedure Language. In
        In *Programming Symposium: Proceedings, Colloque sur la Programmation,* B.
        Robinet, Ed., Springer-Verlag, 1974, pp. 362-376.

[15]   Dennis, J.B.
        "Data Flow Supercomputers".
        Computer 13, 11 (November 1980), 48-56.

[16]   Eckert, J.P., J.C. Chu, A.B. Tonik & W.F. Schmitt
        Design of UNIVAC-LARC System: 1.
        Proceedings of the EJCC, 1959, pp. 59-65.

[17]   Edler, J., A. Gottlieb, C.P. Kruskal, K.P. McAuliffe, L. Rudolph, M. Snir, P.J. Teller
        & J. Wilson
        Issues Related to MIMD Shared-Memory Computers: The NYU Ultracomputer
        Approach.
        Proceedings of the 12th Annual International Symposium On Computer Architec-
        ture, Boston, June, 1985, pp. 126-135.

[18]   Ellis, J.R.
        Culldog: a Compiler for VLIW Architectures.
        The MIT Press, 1986.

[19]   Fisher, J.A.
       Very Long Instruction Word Architectures and the ELI-512.
       Proc. of the 10th, Internation Symposium on Computer Architecture, IEEE Com-
       puter Society, June, 1983.

[20]   Gajski, D.D. & J-K. Peir
       "Essential Issues in Multiprocessor Systems".
       Computer 18, 6 (June 1985), 9-27.

[21]   Gurd, J.R., C.C. Kirkham, and I. Watson
       "The Manchester Prototype Dataflow Computer".
       Communications of ACM 28, 1 (January 1985), 34-52.

[22]   Hennessey, J.L.
       "VLSI Processor Architecture".
       IEEE Transactions on Computers C-33, 12 (December 1984), 1221-1246.

[23]   Hiraki, K., S. Sekiguchi, and T. Shimada
       System Architecture of a Dataflow Supercomputer.
       Computer Systems Division, Electrotechnical Laboratory, Japan, 1987.

[24]   Iannucci, R.A.
       A Dataflow / von Neuamnn Hybrid Architecture.
       Ph.D.Th.Dept. of Electrical Engineering and Computer Science, MIT, Cambridge,
       Mass., (in preparation) 1987.

[25]   Jordan, H.F.
       Performance Measurement on HEP- A Pipelined MIMD Computer.
       Proceedings of the 10th Annual International Symposium On Computer Architec-
       ture, Stockholm, Sweden, June, 1983, pp. 207-212.

[26]   Kuck, D.E. Davidson, D. Lawrie, and A. Sameh
       "Parallel Supercomputing Today and the Cedar Approach".
       Science Magazine 231 (February 1986), 967-974.

[27]   Lampson, B.W. and K.A. Pier
       A Processor for a High-Performance Personal Computer.
       Xerox Palo Alto Research Center, January, 1981.

[28]   Li, Z. and W. Abu-Sufah
       A Technique for Reducing Synchronization Overhead in Large Scale Multi-
       processors.
       Proc. of the 12th, International Symposium on Computer Architecture, June, 1985,
       pp. 284-291.

[29]   Moon, D.A.
       Architecture of the Symbolics 3600.
       Proceedings of the 12th Annual International Symposium On Computer Architec-
       ture , Boston, June, 1985, pp. 76-83.

[30]   Nikhil, R.S., K. Pingali, and Arvind
       Id Nouveau.
       Computation Structures Group Memo 265, Laboratory for Computer Science, MIT,
       Cambridge, Mass., Cambridge, MA 02139, July, 1986.

[31]   Papadopoulos, G.M.
       Implementation of a General Purpose Dataflow Multiprocessor.
       Ph.D.Th., Dept. of Electrical Engineering and Computer Science, MIT, Cambridge,
       Mass., (in preparation) 1987.

[32]   Paterson, D.A.
       "Reduced Instruction Set Computers".
       Communications of ACM 28, 1 (January 1985), 8-21.

[33]   Pfister, G.F., W.C.Brantley, D.A. George, S.L. Harvey, W.J.   Kleinfelder, K.P.
       McAuliffe, E.A. Melton, V.A. Norton, and J. Weiss
       The IBM Research Parallel Processor Prototype (RP3): Introduction and Architec-
       ture.
       Proceedings of the 1985 International Conference on Parallel Processing, Institute
       of Electrical and Electronics Engineers, Piscataway, N.J., 08854, August, 1985, pp.
       764-771.

[34]   Radin, G.
       The 801 Minicomputer.
       Proceedings of the Symposium on Architectural Support for Programming Lan-
       guages and Operating Systems, ACM, March, 1982.

[35]   Rau, B., D. Glaeser, and E. Greenwalt
       Architectural Support for the Efficient Generation of Code for Horizontal Architec-
       tures.
       Proceedings of the Symposium on Architectural Support for Programming Lan-
       guages and Operating Systems, March, 1982.   Same as Computer Architecture
       News 10,2 and SIGPLAN Notices 17,4.

[36]   Rettberg, R., C. Wyman, D. Hunt, M. Hoffmann, P. Carvey, B. Hyde, W. Clark, and
       M. Kraley
       Development of a Voice Funnel System: Design Report.
       4098, Bolt Beranek and Newman Inc., August, 1979.

[37]   Russell, R.M.
       "The CRAY-1 Computer System".
       Communications of ACM 21, 1 (January 1978), 63-72.

[38]   Seitz, C.M.
       "The Cosmic Cube".
       Communications of ACM 21, 1 (January 1985), 22-33.

[39]   Smith, B.J.
       A Pipelined, Shared Resource MIMD Computer.
       Proceedings of the 1978 International Conference on Parallel Proceeding, 1978, pp.
       6-8.

[40]   Thomton, J.E.
       Parallel Operations in the Control Data 6600.
       Proceedings of the SJCC, 1964, pp. 33-39.

[41]   Traub, K.R.
       A Compiler for the MIT Tagged-Token Dataflow Architecture - S.M. Thesis.
       Technical Report 370, Laboratory for Computer Science, MIT, Cambridge, Mass.,
       Cambridge, MA 02139, AUGUST, 1986.

[42]   ALTO:
       A Personal Computer System - Hardware Manual.
       Xerox Palo Alto Research Center, Palo Alto, California, 94304, 1979.

Part III - Software and Languages

Parallel Programming Support in ParaScope

David Callahan
Keith D. Cooper
Robert T. Hood
Ken Kennedy
Linda M. Torczon
Scott K. Warren

Department of Computer Science
Rice University
Houston, Texas 77251-1892

1. Introduction

It now seems clear that almost every new supercomputer design will employ some form of parallelism, because it offers the promise of higher execution speeds at reasonable costs. However, the burden of achieving that promise has been shifted to the programmer, who must decompose his problem into pieces that can be executed in parallel and ensure that the parallel processors assigned to these pieces are properly synchronized to produce a correct answer deterministically. This is not an easy task, because it requires not only an understanding of the problem being solved, but also of the underlying computer architecture and of the data-flow patterns in the program used to solve the problem. Clearly, sophisticated programming support tools are needed to assist in this process.

Typically, the process of parallel programming maps into four distinct tasks.

- *Decomposition.* The programmer must decompose the application problem into parallel tasks. This is most effectively done when the programmer can deal with the problem at a high level of abstraction.

- *Implementation.* The programmer must convert the abstract decomposition into a parallel program in some computer language. This task tends to be highly error-prone because it is fairly difficult to produce a program that precisely implements the vision of the parallel decomposition without making clerical errors.

- *Compilation.* The program must be translated into a machine-language version that can be executed on a parallel processor. This is typically the responsibility of the compiling system, although the sophistication of the compiler determines the extent to which the computer language can hide the details of parallel programming from the user.

- *Debugging.* The programmer, with the help of any available tools, must eliminate any logic errors that are discovered while testing the program. Parallel programming has given rise to a new class of bug: the *schedule-dependent* error. Schedule-dependent errors are difficult to locate because they are manifested only when certain schedules are employed in the parallel region. Frequently, the error disappears during debugging only to reappear when the program is put into production.

If we are to make parallel programming tractable, we must develop better tools in each of these areas.

ParaScope is a parallel programming environment under development at Rice University. It is designed to support the implementation, compilation and debugging phases in an integrated support system. In addition, it will also provide limited support for decomposition. When complete, it will consist of a collection of tools for elaborating a parallel program in an extended dialect of FORTRAN. The tools will include a sophisticated editor for FORTRAN source, an editor for defining whole programs, a powerful compilation system that attempts to enhance the parallelism in the FORTRAN program through sophisticated transformation techniques, and an execution monitor that supports powerful techniques for debugging parallel programs. In addition, the source editor will have facilities to assist the programmer in decomposing a program for parallelism and to find clerical errors in a proposed decomposition.

ParaScope is the logical outgrowth of our research on scientific programming environments and our investigation of techniques for automatic detection of parallelism. Over the past four years, the *setr*[n] project at Rice University has been constructing a sophisticated programming environment for FORTRAN [HoKe85, CoKT86c, ChHo 87]. The current system consists of an integrated collection of tools to aid programming teams in entering and testing FORTRAN programs. It provides sophisticated facilities for managing all the code in the various components of the program. These tools provide the environment with enough control over the source of a program to attempt significant program optimizations that are not possible within a conventional compiler [CoKe 84, Coop 85, CCKT 86, CoKT 86b].

At the same time, the PFC project has been exploring methods for vectorization and automatic detection of parallelism [AlKe 84, AlKe 85, AICK 87]. A significant subproject has developed PTOOL, a sophisticated browser that helps the user find parallelism inherent in an existing program [ABKP 86].

By combining technology developed in these projects, we expect to synthesize a useful tool for parallel programming. The central theme of this work is to develop ParaScope into an environment that runs on a high-performance workstation and assists in the preparation of programs intended for execution on a parallel-vector supercomputer like the IBM 3090 with vector feature.

The paper begins in Section 2 with a description of our past research on programming environments and automatic detection of parallelism. Section 3 describes our plans for ParaScope, including the six main research areas:

- incremental dependence analysis in a structure editor,
- improvements to the existing interprocedural analysis,
- techniques for debugging parallel programs,
- whole program planning for parallelization,
- code generation for parallel machines, and
- the display of complex program annotations.

Finally, Section 4 contains some concluding remarks.

2. Background

2.1 The $\mathbb{R}^n$ FORTRAN Programming Environment

The $\mathbb{R}^n$ programming environment is an integrated collection of tools designed to assist programmers building numerical software systems in FORTRAN [HoKe 85, CoKT 86c]. The environment provides balanced support for the programming process, incorporating tools to aid in the construction of whole programs as well as individual modules. Along these lines, it provides editors for the source code of a single routine, called a *module*, and for the structural description of a program, called a *composition*. It has two compilers, a *module compiler* that deals with intraprocedural optimization and code generation and a *program compiler* that deals with interprocedural issues. The $\mathbb{R}^n$ *execution monitor* supports execution of hybrid programs in which some modules are interpreted and others are executed from compiled code. All of the tools interact with the user under the control of a window-based *monitor*. Information is passed among the various tools by saving it in a central database. A brief description of the principal components follows:

Module Editor

> The module editor combines a knowledge of FORTRAN with access to the database to help the user construct syntactically correct programs. It provides a convenient blend of structure editing and text editing, allowing the user to shift freely between these two paradigms. A module is defined by a set of entry point specifications; the different versions of a module all implement the same entry points. The editor directly constructs an *abstract syntax tree* representation of the module; this internal form for the program is used throughout the environment.

Composition Editor

> In $\mathbb{R}^n$, both programs and composite modules are represented by their structural descriptions, called compositions [CKTW 86]. Compositions have hierarchical structure; they contain individual modules and other compositions. Import and export lists associated with each level of the composition provide a scoping and renaming facility. In essence, the $\mathbb{R}^n$ notion of a composition is itself a module interconnection language, albeit one with a screen-oriented concrete syntax. The user creates and modifies these descriptions with the composition editor, a structure editor for this interconnection language. The editor checks the program for consistency and completeness. It also ensures that entry point specifications match the call sites to which they are bound.

Program Compiler

> The program compiler's task is to construct an executable image of a program that is both consistent with the current state of the source code for its components and fully optimized. When invoked on a specific program, it checks the records updated by the module editor and the composition editor to determine what changes have occurred since the last compilation. It uses this information to update the interprocedural summary, aliasing, and constant propagation information available for the program. Next, it performs a recompilation analysis to construct a complete list of procedures that must be

recompiled. Finally, it looks at the set of procedures being recompiled to determine where interprocedural optimizations like linkage tailoring are likely to be profitable. Given all this information, it then invokes the module compiler to perform the needed recompilations [CoKT 86a, CKTW 86].

Module Compiler

The module compiler's task is to translate one or more modules into optimized code for the target machine, using interprocedural information and optimization directives provided by the module compiler. Essentially, the module compiler is an optimizing code generator that makes use of the information collected in the program compiler. It has been carefully structured to allow experimentation with different optimization and recompilation strategies.

Execution Monitor

The execution monitor allows the programmer to execute a program constructed by the environment. It provides the user with a range of instrumentation levels, including execution of hybrid programs in which some modules are run from compiled, optimized code and other modules are run interpretively [ChHo 87]. Instrumentation levels are dynamic; the programmer can change them whenever execution is paused at a breakpoint. Breakpoints can be specified in terms of either the program's composition or the source code for a specific module. We plan to support reversible execution; efficient implementation of this feature will rely on using the flow-insensitive interprocedural summary information computed by the rest of the environment.[1]*

The $\mathbb{R}^n$ programming environment has grown to approximately 105,000 lines of C and has itself become a significant experimental resource. It is designed in a modular fashion to permit rapid prototyping of experimental tools by providing them with a layer of environment services such as screen and database management. We expect to distribute an experimental version of the environment in late 1987.

2.2 PFC and PTOOL

Since 1978, Rice has been conducting an active program of research in software for vector and parallel supercomputers. PFC (the Parallel FORTRAN Converter) is a system that automatically vectorizes FORTRAN programs by performing a sophisticated analysis of *dependences*. A dependence exists between two statements if one statement can store into a location that is later accessed by the second statement. Although most optimizing compilers analyze dependences in a program, they use a particularly naive treatment of arrays. Vectorization systems employ a much more powerful analysis that is fairly effective in dealing with subscripted references in loops [Kuck 78, KKLW 80, AlKe 86]. Research on PFC has

---

[1]   This information is used when an interpreted subroutine makes a call to compiled code. Without the side-effect information, the interpreter would need to checkpoint the entire memory state accessible outside the calling routine both before and after the call and then compare the two images to determine what variables were changed by the call. With the information, it need only examine those variables contained in the interprocedural MOD set for the call site, a much less formidable task.

concentrated on finding vectorization algorithms efficient enough for use in a compiler. An indication of its success is that PFC served as the prototype for the IBM VS FORTRAN Version 2 vectorizing compiler for the 3090 Vector Feature [ScKo 86], which achieves excellent results while remaining reasonably efficient.[2])

Since completion of the original PFC in 1982, research has continued in three areas. First, algorithms to perform interprocedural analysis on whole programs have been added. The approach used is based upon results from the batch version of the algorithm developed for $\mathbb{R}^n$. The aim of this work is to determine the impact of global program knowledge on vectorization and parallelization. Currently, PFC analyzes side effects of procedure calls, aliasing patterns, and constants propagated across procedure boundaries.

Second, we have been examining the use of PFC's analysis phase in parallel programming tools. Our first effort, called PTOOL, is an interactive adviser designed to assist in the prevention of errors arising from unintentional data sharing or unforeseen load-store orders for shared data in parallel programs [ABKP 86]. PTOOL is really a sophisticated browser for a database of interstatement dependences created by PFC. It permits the user to select a loop in a sequential FORTRAN program and ask whether or not the iterations may be run in parallel. If the answer is "no", PTOOL will display all the dependences that prevent parallelization.

While PTOOL is extremely helpful for identifying loops that can be parallelized, it provides no assistance in the generation of parallel code. A third project is investigating automatic generation of code for multiprocessors. This has led to another derivative of PFC, called PFC Plus, that not only recognizes loops whose iterations can be run in parallel but also performs sophisticated transformations, such as loop interchange, loop alignment, and code replication to enhance the parallelism available [AlKe 85].

PFC now contains approximately 95,000 lines of PL/I code and runs on an IBM 4341. PTOOL is already in use at Los Alamos National Laboratory and is scheduled to be installed at Livermore and Argonne. Both PTOOL and PFC Plus will be installed on the Cornell Theory Center's IBM 3090 supercomputer system. These systems represent a significant research resource because they are relatively easy to modify. This makes it possible to add and evaluate new transformation methods rapidly. In addition, PTOOL has value as an educational tool. It is now used to teach compiler students about dependences in programs.

3. ParaScope

It is the goal of the ParaScope project to produce a programming environment that will assist in the formulation, implementation, and debugging of parallel FORTRAN programs.

---

[2]) Although the IBM compiler is based on an early version of the PFC system, it is quite effective at vectorization and runs only 25% slower than the scalar VS FORTRAN compiler with optimization.

In the environment we envision, a programmer will prepare a FORTRAN program (containing only standard FORTRAN constructs) using the environment's tools. The program might consist entirely of new code or it might incorporate packaged source modules. When the programmer is ready to consider the implications of parallelism, the system will bring to bear the analytical tools from the compiling system, reporting back on its successes and failures.

For those regions of the program that the compiler cannot run in parallel, the programmer can then examine the dependence information to see if a simple revision might permit more parallelism. At each step, the system will report success or failure and permit further investigation of dependences. When the programmer is satisfied, he can instruct the environment to generate code for the target machine and run the program.

Under this approach, program development is a cooperative effort by both the programmer and the system, exploiting the strengths of each. The system performs the tedious analysis of dependences, freeing the programmer to develop more highly parallel algorithms. Given feedback from the system, the programmer should be able to experiment with innovative approaches.

If successful, this project will produce a sophisticated programming system that not only helps the programmer produce correct, well organized code, but also assists him in tailoring that code to achieve high performance on parallel architectures. The following sections present details on the six subprojects we envision.

3.1 Dependence Analysis in a Structure Editor

In the automatic detection of parallelism, the key analytical information used by a compiler is a *dependence graph* [KKLW 80, Kenn 80, KKLP 81, Alle 83] for the procedure. To create a responsive tool for parallel programming, we will need to perform dependence analysis on the procedure as it is modified in the environment's structure editor for FORTRAN. While structure editing systems that incrementally update semantic information are reasonably common, we are aware of no system that maintains a consistent dependence graph of sufficient complexity to support the transformations required for parallel code generation.

Our experience with the PTOOL system has convinced us of the value of presenting dependence based information to programmers trying to understand why particular loops will not run in parallel. Unfortunately, because PTOOL relies on a batch analysis of the program, users have found the delay between typing a proposed change into the source and getting feedback about its impact to be frustrating. Constructing a useful, on-line parallel programming advisor will require a method for rapidly updating the dependence information for a procedure in response to one or more editing changes.

Given the need for incremental dependence analysis, a natural starting point for the research is to consider application of a general technique like an attribute grammar framework [Reps 82] or one of the existing incremental data-flow analysis frameworks [Zade 84, Ryde 83, CaRy 86]. Unfortunately, the computation of dependence analysis appears to be sufficiently complex to make re-casting it in an attribute grammar framework unbearably inefficient. Similarly, there appears to be no natural way to cast the problem in a data-flow analysis framework. Certainly, the resulting framework would be so complex that the standard algorithms from global data-flow analysis would not achieve their fast time bounds.

We believe that the speed of incremental updates to this information is critical to the responsiveness of the overall parallelism advisor. To achieve the goals of this project, we intend to develop efficient incremental updating techniques to deal with both factual and structural changes.

3.2 Improvements to Interprocedural Analysis

Our experience with interprocedural data-flow analysis in both $\mathbb{R}^n$ and PFC has convinced us that improvements are needed in both the theory and the implementation. Two particular problems present themselves: the need for more precise treatment of arrays and the need for incremental updating techniques. The next two subsections provide more detail on these problems.

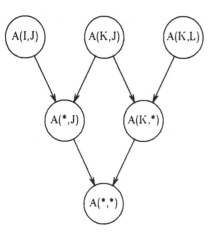

Figure 1.   Simple side-effect lattice.

3.2.1 Regular Section Analysis

Our experience with using interprocedural summary information in a working system for detecting parallelism has shown that the granularity of conventional summary information is too coarse to allow effective detection of parallelism in loops that contain call sites. The problem is that the current analysis treats whole arrays as single units. Thus, it is able to determine whether an array is modified somewhere, but not whether it is modified in only a single column or row. This limitation is disastrous for parallelization because the most effective way to parallelize a loop is through data decomposition, in which each parallel iteration works on a different subsection of a given array.

Hence, some mechanism for determining the subsections actually affected by interprocedural side effects is needed. Triolet has proposed a method that finds the convex hull of the set of array locations affected as a side effect of a procedure call [TrlF 86]. Unfortunately, this method is too expensive to use in a compiler. Therefore, we seek to achieve a more limited goal: to recognize some important special cases of array side effects. For example, it would be extremely useful if we were simply able to recognize when the modification of an array by a procedure call is limited to a single column or row of the array.

Fortunately, a generalization of the approach currently used to solve interprocedural data-flow analysis problems can be used to develop more precise information about side effects. Although it may not be immediately clear from perusing the published papers [CoKe 84, CoKT 86c], the Cooper-Kennedy algorithm for summary problems can be extended to work on lattices. Consider the example lattice of reference patterns to the array A shown in Figure 1 on page 97. Note that I, J, and K are arbitrary symbolic input parameters to the call. To incorporate a more accurate treatment of arrays, we must extend the side effect analysis to use more general vectors of lattice elements rather than simple bit vectors.

Our technique for computing summary information is particularly well suited to this type of analysis because it divides the computation of interprocedural information between a step that traces the impact of chains of parameter binding and a step that uses this binding map to transform local information into the desired interprocedural information [CoKe 84]. Applying the method to a more complex domain of interprocedural facts doesn't change the first part of the analysis; only the second step requires modification. Since the first step requires time that is nearly linear in the size of the program, the cost of performing the meet operations in the second step may dominate the cost of the entire analysis. This makes choosing an appropriate lattice important, not only from the perspective of modeling facts that aid in the parallelization, but also because the lattice will directly impact the cost of performing the analysis.

In his dissertation, Callahan proposes several regular section lattices [Call 87]. As part of this research project, we will implement a version of the interprocedural summary algorithm that can operate on such lattices and use this to determine the efficacy of his various models. We believe that this type of analysis will significantly enhance our ability to detect parallelism.

3.2.2 Incremental Update Techniques

The current version of the $\mathbb{R}^n$ environment completely recomputes interprocedural information each time a program is compiled. The analysis is sufficiently fast that this has not been a major problem in the past. However, we expect that regular section analysis will require substantially more computing resources than the current techniques. This will increase the desirability of incremental methods for updating interprocedural information in response to a program change. We have identified several promising approaches for these updates [Coop 83, CoKe 84]. In implementing an update technique, we must examine not only our own work, but also the ongoing research into incremental updating techniques both at Rice and elsewhere [Reps 82, Zade 84, CaRy 86]. We plan to pick the most promising technique emerging from ongoing research on incremental analysis, implement it in ParaScope, and evaluate its efficacy. In performing this research, it will be important to evaluate the tradeoff between incremental and parallel evaluation methods. It seems likely that batch style algorithms may be more amenable than incremental techniques to parallel execution.

3.3 Parallel Debugging

Currently, the $\mathbb{R}^n$ execution monitor (ExMon) supports debugging a sequential program on the local machine. In ParaScope, we will enhance it to support debugging programs executing on remote machines and to support debugging parallel programs. These extensions divide into two fundamentally different tasks: providing the mechanisms to control a process on a remote machine and examine or change its internal state, and developing paradigms to allow the programmer to relate source code constructs to the execution state of the parallel program. Both of these are necessary before a practical $\mathbb{R}^n$-style execution monitor can be constructed for parallel programs.

3.3.1 Implementing Remote Debugging

Fortunately, the implementation of ExMon was designed in a manner that should facilitate extending it to deal with processes on remote machines. It is implemented as two UNIX processes. The debugger itself, as well as all interpretive execution, runs in a foreground process. Whenever, compiled code is running, it runs in a separate process, with its own address space, in the background. Consistency of data values is insured by always keeping the correct values in the address space of the compiled process, using the UNIX system call *ptrace*.

To implement a version of ExMon that supports remote debugging will require implementation of a *remote debugging kernel* that supports the following operations:

- a combination of *fork*(2) and *execve*(2)

    The kernel must be able to start execution of a given file on the remote machine.

- *ptrace*(2)

    In order to read and write in the address space of the target program and to set breakpoints there, the remote kernel must provide a function equivalent to the UNIX *ptrace* system call.

- pause process

    The kernel needs some mechanism that will allow the debugger to pause a running target program.

- *wait*(2), *wait3*(2)

    In order to determine when the target program has stopped at a breakpoint, the remote kernel must support something similar to the UNIX *wait* and *wait3 calls*. To avoid polling, the remote machine should asynchronously advise the debugger of a change of state in the child.

- get symbol table from executable

    In order to determine the location of various identifiers in the target program, the remote kernel must support reading the target program's symbol table.

- stack abstraction

The kernel must provide operations that permit the run-time stack to be manipulated abstractly. This includes routines for addressing the various storage classes of identifiers as well as routines for pushing and popping stack frames.

- redirection of *stdin, stdout, stderr*

  In order to enter input data from the debugger's machine and to get the target program's output on it, the remote kernel must support some form of redirection of standard I/O.

An implementation of NFS on both the local and remote machine would simplify several problems, notably reading files on the remote machine and examining the target program's symbol table. Parallel programs involving multiple processes can be handled by allowing the debugger to manage more than one background process. Thus, with a remote debugging kernel supporting these operations via a remote procedure call mechanism, we expect to use ExMon running on a workstation to debug programs running on a remote supercomputer.

3.3.2 Relating State to Source

Most manufacturers are extending their sequential source debuggers to parallel systems by permitting the user to halt processes and single-step each process. Unfortunately, this approach doesn't provide a practical tool for debugging complex parallel programs. We intend to attack two of the problems that arise in dealing with the execution state of a parallel program: relating that state back to the user's source code in a meaningful way and providing tools to help the programmer cope with the nondeterministic nature of execution.

The first problem arises regularly in attempts to understand optimized code. The problem is made more acute by the radical transformations that a compiler attempts to discover additional parallelism. ParaScope will help the programmer understand the relationship between source and machine code by producing compiler generated annotations that can be examined in the source editor using the facilities discussed in Section 3.6. One possible form for the annotations would be a transformed version of the source, in a pseudo-language that exposes constructs hidden in the original FORTRAN source. Other techniques will surely suggest themselves as the work on the display of annotations matures.

The second problem may well be the most difficult problem in debugging code for shared memory multiprocessors. Simply put, it is very hard to recreate with a debugger the sequence of events that leads to an error of unintentional data sharing. Although techniques based upon non-intrusive tracing show great promise [GeHo 83], the hardware is not yet efficient enough to routinely support this approach. In ParaScope, we will use the information discovered during analysis and compilation to provide clues to the location of errors at run time.

For example, if an incorrect value is detected in a parallel program at a point where the sequentially scheduled version produced the correct value, the debugging system would be invoked. It could trace back along dependence edges to locations inside parallel regions that are potential sources of the problem. Then, it could use *adversary scheduling*, a technique that employs dependence analysis to pick schedules likely to lead to errors, to locate the error. If the computation inside the parallel region was involved in a dependence that the compiler ignored under the programmer's direction, the system could generate a processor schedule that caused the dependence to be violated. In other words, the debugger would step the processors in an order that is most likely to give rise to the value-passing pattern that caused static analysis to suggest that the sequential and parallel semantics differed.

3.4 Whole Program Planning

In a system for automatically decomposing a large program for parallel execution, optimizing transformations should be planned from a global perspective. For example, when choosing the loop to run in parallel from among several possibilities, it is desirable to select the outermost loop in order to generate less frequent synchronizations. However, if the program consists of many separately-compiled subroutines, the compiler will have trouble determining if a given loop is actually the outermost loop. Suppose that it confronts a loop within a subroutine and that the version of Parallel FORTRAN does not allow nested parallel DO's. In this case, the compiler should only select the loop if the subroutine is not called from within a parallel loop in some other routine. Determining when this is the case requires information about the whole program.

As a second example, consider inline substitution. When properly used, it can uncover parallelism that is very hard to see when subroutines are examined separately. Furthermore, it permits code from the substituted procedure to be tailored to the situation at the point of call, making more optimizations possible. This combination of effects can be extremely beneficial. However, unrestricted use of inline substitution can lead to an explosion in code size.

To make appropriate use of inline substitution, the compiling system must plan the optimizing transformations, including parallelization, for the whole program. The approach we envision constructs the call graph, gathers information about each procedure (including information for the inline substitution phase), computes interprocedural data-flow information and performs inline substitution. It then determines the important environmental information, like whether or not a call to the routine is contained in a parallel loop, to pass to the compiler for each individual compilation. Finally, it invokes compilations of the individual modules, passing the environmental data to the compiler.

The implications of whole program planning in a programming environment are extremely complex. In ParaScope, planning will be the responsibility of the program compiler. To make the system work efficiently, we will need to develop incremental methods for revising plans in response to change, similar to the methods we developed for limiting recompilation [CoKT 86b, Torc 85]. An important preliminary step to such algorithms is to implement recompilation limitation and gain experience with it.

The program planning scheme also appears to provide a natural place to identify larger granularity parallelism. It seems likely that looking for large grain parallelism on a whole program basis can lead to more effective use of a multiple processor machine by a single FORTRAN program. By identifying individual routines or groups of routines whose executions can be run in parallel, the program compiler should be able to assign these tasks to separate processors.

An important application of this research is to develop techniques for planning the optimizations of whole programs that are being prepared for execution on background supercomputers. ParaScope will include a powerful new program compiler that will manage programs on a background mainframe. The enhanced techniques should permit program compilation and module compilation to take place on the mainframe itself. Implementing this scheme will require cooperation with the mainframe FORTRAN compiler to achieve highly optimized whole programs.

3.5 Parallel Code Generation

To complete the task of preparing a program for execution in ParaScope, there must be a compiler that handles all the details of generating code for the supercomputer. To this end, the $\mathbb{R}^n$ optimizing module compiler, currently in prototype form, will be completed and extended to support generation of code for parallel supercomputers. Since the necessary dependence analysis will be performed in the editor and results stored in the database, the compiler can use this information in transforming and optimizing the program without directly paying the price for the analysis. This project would build upon our work on optimization and transformation of programs for parallel execution in PFC [AlKe 85, AICK 87].

The optimizing compiler will benefit in a number of ways from the use of dependences, because they are more precise in their handling of subscripted variables than traditional data-flow information. Cytron has discussed a variety of interesting applications for such dependences [Cytr 86]. Allen and Kennedy [AlKe 86] have already shown that memory traffic for scalar computations can be reduced significantly by use of register allocation techniques based upon dependence. We also expect several other new scalar optimizations to present themselves.

The main problem to be attacked is the generation of parallel code for a supercomputer. Unlike PFC, this work would be in the context of a compiler. Hence, a number of machine-dependent issues can be explored. For example, what is the correct tradeoff between vector and parallel execution? Also, how can memory traffic be minimized for highest performance on a machine with a complex memory hierarchy?

If we attempt to isolate the transformations in one section of the compiler, we should also be able to produce a version of the compiler that generates FORTRAN annotated with parallel execution primitives. This would permit the resulting program to be compiled by the VS FORTRAN compiler for execution on the IBM 3090. If done properly, this goal would require that IBM make extensions to VS FORTRAN to accept the parallel primitives along with the interprocedural information produced by the program compiler.

The ParaScope optimizing module compiler should provide an ideal laboratory for investigating these and other issues related to compilation for parallel supercomputers.

3.6 Display of Complex Program Annotations

Designing a good user interface for ParaScope's sophisticated program analysis and transformation capabilities is a major challenge. $\mathbb{R}^n$ produces and consumes many kinds of information about FORTRAN programs in addition to source code;

that information is voluminous, diverse in structure, and highly inter-related. Parallel programming features will add to this information considerably. The current version of $\mathbb{R}^n$ uses a graphical window system to display each kind of information in a separate window, with few provisions for displaying inter-relationships. Up to now, the study of user interface issues has not been an explicit goal of the $\mathbb{R}^n$ project.

If ParaScope is to become a useful tool for parallel programming, we need to develop mechanisms to help the ParaScope user deal with the complex web of information surrounding his FORTRAN program. Our approach is to view the information as a *document*, an "electronic book" written in the dynamic medium of computer graphics instead of on paper [KayG 77, YMvD 85, WeyB 85]. We think of all the extra information as *annotations* of the FORTRAN source text, analogous to a book's footnotes, figures, and appendices. Unlike a paper book, portions of the annotated FORTRAN can be selectively revealed or concealed, viewed in a choice of different formats, or edited. There has been much research on such dynamic media, but little of it has dealt with computer programs and none that we know of has focused on the complex, inter-related information managed by a programming environment.

3.7 Annotations

Some of the information managed by ParaScope comes from the programmer, and some is produced by ParaScope as a product of program analysis or transformation. The programmer provides four kinds of information to ParaScope. *Project management information* includes ownership of versions and revision histories of modules. *Programming information* includes formal specifications, design documentation, source code modules, and compositions of modules into programs. *Compiling information* includes such choices as where to perform in-line substitutions, what target machine to assume, and which modules to run interpretively. *Testing information* includes input test data, expected output, and debugging session logs.

ParaScope provides information to the programmer as a result of analyzing, transforming, or executing FORTRAN source. Conventional analysis finds syntactic and semantic errors in FORTRAN source code, connects declarations and uses of identifiers, and determines variables aliased by common or equivalence statements. Data-flow analysis finds constant valued expressions, connects stores with loads, and determines what dependences exist between statements or expressions. Transformation exposes parallelism at the source code level. In addition to producing normal program output, execution can produce other important information: suspended execution contexts, traces of adversary schedules, and performance measurements.

3.8 Source Code Enhancement

To display these annotations, we will explore techniques from four areas of related research: structured documents [DKLM 84, YMvD 85], hypertext [Carm 69, Nels 74], Knuth's WEB system of structured documentation [Knut 84], and graphical programming [Raed85, Reis84, MorH85]. We will begin by applying some of these techniques in isolation, using the current $\mathbb{R}^n$ system as a test bed. Our eventual aim is to find ways of unifying these techniques into a simple but powerful model for annotated FORTRAN.

In the first phase of research, we will use the existing $\mathbb{R}^n$ user interface as a test bed to explore ways of presenting annotations. Later, we will build an entirely new user interface based on the unifying notion of annotated documents. One benefit of this research would be to ensure that ParaScope is easy to use when it is finished.

4. Conclusions

The first vector supercomputers appeared on the market in the early to mid seventies. Yet, because of the lag in developing supporting software, it is only recently that vectorizing compilers powerful enough to effectively utilize vector hardware have been developed.

Since parallel programming is a much more complex task than vectorization, we expect the challenge of producing adequate programming support to be much greater. In the ParaScope project, we will be exploring the leverage to be gained through an integrated collection of tools in which each tool depends on the others for important information. For example, the editor will depend on the interprocedural analyzer, which itself depends on the results of editing other modules. The debugger uses dependence information to assist in the location of potential problems. The user interface permits abstract displays of the data-flow information within a program.

We believe that it is essential to have this sort of cooperation to provide adequate support for programming on the evolving class of highly parallel machines.

5. References

[Alle 83]   J.R. Allen, "Dependence analysis for subscripted variables and its application to program transformations," Ph.D. dissertation, Department of Mathematical Sciences, Rice University, Houston, Texas, April, 1983.

[ABKP 86]   J.R. Allen, D. Baumgartner, K. Kennedy, and A. Porterfield, "PTOOL: a semi-automatic parallel programming assistant", *Proc. 1986 International Conference on Parallel Processing*, IEEE Computer Society Press, Washington, D.C., 1986.

[AICK 87]   J.R. Allen, D. Callahan and K. Kennedy, "Automatic decomposition of scientific programs for parallel execution,"*Conf. Record, 14th POPL*, January, 1987.

[AlKe 84]   J.R. Allen and K. Kennedy, "PFC: a program to convert FORTRAN to parallel form," *Supercomputers: Design and Applications*, K. Hwang, ed., IEEE Computer Society Press, Silver Spring, MD, 1984, 186-203.

[AlKe85]   J.R. Allen and K. Kennedy, "A parallel programming environment," *IEEE Software 2(4)*, July 1985.

[AlKe 86]   J.R. Allen and K. Kennedy, "Vector register allocation", Technical Report 86-45, Rice University, Department of Computer Science, December, 1986.

[CCKT 86]   D. Callahan, K.D. Cooper, K. Kennedy, and L. Torczon, "Interprocedural constant propagation", *Proc. SIGPLAN '86 Symposium on Compiler Construction, SIGPLAN Notices,* 21(7), July, 1986.

[Call 87]   D. Callahan, "A global approach to the detection of parallelism", Ph.D. Thesis, Department of Computer Science, Rice University, February, 1987.

[Carm 69]   " S. Carmody et al., "A hypertext editing system for the /360", *Pertinent Concepts in Computer Graphics,* M. Faiman and J. Nievergelt (eds.), Univ. of Illinois Press, 1969, 291-330..

[CaRy 86]   M. Carroll and B.G. Ryder, "An incremental algorithm for software analysis", *Proceedings of the ACM SIGSOFT/SIGPLAN Software Engineering Symposium on Practical Software Development Environments, SIGPLAN Notices* 22(1), January, 1987.

[ChHo 87]   B.B. Chase and R.T. Hood, "Selective interpretation as a technique for debugging computationally intensive programs", *Proc. SIGPLAN '87 Symposium on Interpreters and Interpretive Techniques,* June, 1987.

[Coop 83]   K.D. Cooper, "Interprocedural Data Flow Analysis in a Programming Environment," Ph.D. Dissertation, Department of Mathematical Sciences, Rice University, May 1983.

[Coop 85]   K.D. Cooper, "Analyzing aliases of reference formal parameters" *Conf. Record, 12th POPL,* January, 1985.

[CoKe 84]   K.D. Cooper and K. Kennedy, "Efficient Computation of Flow Insensitive Interprocedural Summary Information," *Proc. SIGPLAN '84 Symposium on Compiler Construction, SIGPLAN Notices* 19(6), June 1984.

[CoKT 86a] K.D. Cooper, K. Kennedy, and L. Torczon, "Optimization of compiled code in the $\mathbb{R}^n$ programming environment", *Proc. of the 19th Annual Hawaii International Conference on System Sciences,* 1986.

[CoKT 86b] K.D. Cooper, K. Kennedy, and L. Torczon, "Interprocedural Optimization: Eliminating Unnecessary Recompilation", *Proc. SIGPLAN '86 Symposium on Compiler Construction, SIGPLAN Notices,* 21(7), July 1986.

[CoKT 86c] K.D. Cooper, K. Kennedy, and L. Torczon, "The impact of interprocedural analysis and optimization in the $\mathbb{R}^n$ programming environment", *ACM TOPLAS* 8(4), October, 1986.

[CKTW 86]   K.D. Cooper, K. Kennedy, L. Torczon, A. Weingarten, and M. Wolcott, "Editing and compiling whole programs", *Proc. of the Second SIGPLAN/SIGSOFT Symposium on Practical Software Development Environments, SIGPLAN Notices* 22(1), January, 1987.

[Cytr 86]   R. Cytron, "On the implications of parallel languages for compilers", Technical Report RC 11723, IBM T.J. Watson Research Center, Yorktown Heights, N.Y., 1986.

[DKLM 84]   V. Donzeau-Gouge, G. Kahn, B. Lang, and B. Melese, "Document structure and modularity in Mentor", *Proc. SIGSOFT/SIGPLAN Symposium on Practical Software Development Environments,* 1984, 141-148.

[GeHo 83]   W.M. Gentleman and H. Hoeksma, "Hardware assisted high level debugging", *Proc. SIGSOFT/SIGPLAN Symposium on High Level Debugging, SIGPLAN Notices* 18(8), August, 1983.

[HoKe 85] R.T. Hood and K. Kennedy, "A Programming Environment for FORTRAN", *Proc. of the 18th Annual Hawaii International Conference on System Sciences,* 1985.

[KKLP 81] D.J. Kuck, R.H. Kuhn, B. Leasure, D.A. Padua, and M. Wolfe, "Compiler transformation of dependence graphs," *Conf. Record, 8th POPL,* January, 1981.

[KKLW 80] D.J. Kuck, R.H. Kuhn, B. Leasure, and M. Wolfe, "The structure of an advanced vectorizer for pipelined processors," *Proc. IEEE Computer Society Fourth International Computer Software and Applications Conf.,* IEEE, October 1980.

[KayG 77] A. Kay and A. Goldberg, "Personal dynamic media", *IEEE Computer,* 10(3), 1977.

[Kenn 80] K. Kennedy, "Automatic translation of FORTRAN programs to vector form," Rice Technical Report 476-029-4, Rice University, October 1980.

[Knut 84] D.E. Knuth, "Literate programming", *Computer Journal,* 10(3), 1977, 31-41.

[Kuck 78] D.J. Kuck, *The Structure of Computers and Computations, Vol. 1,* Wiley, New York, NY, 1978.

[MorH 85] M. Moriconi and D.F. Hare, "PegaSys: a system for graphical explanation of program designs", *Proc. SIGPLAN '85 Symposium on Language Issues in Programming Environments,* June 1985, 148-160.

[Nels 74] T.H. Nelson, *Dream Machines.* Published by the author, 1974.

[Raed 85] G. Raeder, "A survey of current graphical programming techniques", *IEEE Computer,* 18(8), 1985, 11-25.

[Reis 84] S.P. Reiss, "PECAN: program development systems that support multiple views", *Proc. Seventh Int'l Conf. Software Engineering,* March 1984, 324-333.

[Reps 82] T. Reps, "Optimal-time incremental semantic analysis for syntax-directed editors" *Conf. Record, 9th POPL,* January, 1982.

[Ryde 83] B.G. Ryder, "Incremental Data Flow Analysis", *Conf. Record, 10th POPL,* January, 1983.

[ScKo 86] R.G. Scarborough and H.G. Kolsky, "A vectorizing FORTRAN compiler," *IBM J. Research and Development* 30(2), March, 1986.

[Torc 85] L. Torczon. "Compilation dependences in an ambitious optimizing compiler," Ph.D. Dissertation, Department of Computer Science, Rice University, Houston, Texas, May, 1985.

[TriF 86] R. Triolet, F. Irigoin, and P. Feautrier, "Direct parallelization of call statements", *Proc. SIGPLAN '86 Symposium on Compiler Construction, SIGPLAN Notices* 21(7), July, 1986.

[WeyB 85] S. Weyer and A. Borning, "A prototype electronic encyclopedia", *ACM Trans. Office Information Systems,* 3(1), 1985, 63-88.

[YMvD 85] N. Yankelovich, N. Meyrowitz, and A. van Dam, "Reading and writing the electronic book", *IEEE Computer,* 18(10), 1985, 15-29.

[Zade 84] F.K. Zadeck, "Incremental data flow analysis in a structured program editor", *Proc. SIGPLAN '84 Symposium on Compiler Construction, SIGPLAN Notices* 19(6), June, 1984.

Automatic Restructuring of Fortran Programs for Parallel Execution

Constantine D. Polychronopoulos

Center for Supercomputing Research and Development
and Department of Electrical and Computer Engineering
University of Illinois at Urbana-Champaign
104 South Wright Street
Urbana, Illinois 61801, USA

Abstract

With the widespread of high performance parallel processor computers, the need
for powerful restructuring compilers and software environments for program
development has become more pressing than ever before. In this paper we discuss the usefulness of automatic program restructuring, and present some well-
known and some new program transformations that can be used to extract parallelism out of serial Fortran programs. Additional functions that could be carried
out by a compiler are also discussed.

1. Introduction

The ever increasing demand for computational power in many areas of science
and engineering is an undeniable fact. In the first few generations of computer
systems, increased computational speed was primarily obtained through advances
in technology that allowed higher degree of integration and faster switching circuits. Even though technology still advances allowing us to pack more circuitry per
unit of surface and decrease transmission delays, another approach for increasing
performance has been the focus in the last decade. This approach focuses on
parallel architectures and software. Through the replication of computational elements that are interconnected in some regular structure, programs can execute
on multiple processors and access multiple memory banks. In other words computations (and memory transfers) can be performed in parallel.

Parallelism is an old concept and was applied to a limited degree in some of the
early computer systems. First in the form of I/O channels that allowed overlapped
computation and I/O, and later in the form of multiple functional units that could
operate in parallel. Technology constraints and software limitations however did
not make it feasible to design and build parallel machines until later in the 70's.

Informally we could say that parallelism is the result of the concurrent operation
of different components of a computer system. Similarly, a machine can be char-

---

This work was supported in part by the **National Science Foundation** under Grant No. NSF
DCR84-10110 and NSF DCR84-06916, the **U.S. Department of Energy** under Grant No. DOE
DE-FG02-85ER25001, and the **IBM** Donation

acterized as parallel if some or all of its components can operate simultaneously. In our case, parallel and parallelism refer to computer systems that can execute several different instructions or several instances of the same instruction during a single clock period.

Parallel architectures also require more sophisticated compilers and operating systems. One of the major tasks of compilers for parallel machines is to discover and exploit parallelism in a conventional serial program. Such compilers restructure the source program from serial into parallel form and are usually called *restructurers* [3], [4], [8], [16]. The more effective the restructuring compiler, the better the program and system performance. In this paper we describe techniques for program restructuring and address some important issues for parallelizing compilers. We also give a brief introduction to Parafrase II, a parallelizing compiler project underway at the University of Illinois. Some common transformations as well as new transformations are presented.

In Section 2 we briefly discuss vector and parallel machines. Program parallelism, sources of parallelism and granularity issues are the subjects of Section 3 that gives the context for vectorizing and parallelizing compilers. The basic data dependence definitions are given in Section 4. Section 5 discusses automatic program optimization and restructuring techniques. In Section 5.1 we present an overview of some common (and usually machine independent) optimizations. Section 5.2 covers specific vectorization and parallelization techniques. Some new optimizations (cycle shrinking, loop coalescing, and code reordering for reducing communication) are also presented in Section 5.2. In Section 6 we briefly look at an issue that has not attracted the appropriate attention thus far, namely run-time dependence checking and parallelism exploitation. Finally, Section 7 gives some concluding remarks.

2. Vector and Parallel Machines

Several attempts have been made to characterize computer systems based on their instruction sets, control, processor organization, and so on. So far there is no universally acceptable taxonomy. Some of the most commonly used terms to characterize the functionality of computer systems are pipelined, vector, array, and multiprocessor systems. However there is no clear distinction from the instruction set point of view for example, between pipelined or array organizations.

One could say that "vector" refers to computers with vector instructions in their instruction repertoire. Pipelined and array machines fall into this category. Pipelined systems usually have pipelined functional units, as the term implies. Examples include the Cray 1, the CDC Cyber 205, the Fujitsu VP-100/200, the Hitachi S-810, and the Convex-1 computers [22], [23]. Pipelining though can also be used at a higher level. One common activity that is pipelined in most modern computers is the instruction execution cycle. The three different phases (instruction fetch, decode, and execute) can be pipelined by having different instructions going through different phases of execution simultaneously. Array computers usually come with a number of identical ALUs interconnected in some symmetric structure (e.g., linear array, mesh, ring). Some existing array machines include the Goodyear MPP, ICL DAP, Illiac IV, and the Connection machine. Pipelined and array systems are also characterized by their single control unit. According to Flynn's taxonomy these are single-instruction-multiple-data or SIMD architectures. They are synchronous systems in the sense that the control unit and all computational elements are driven by a common clock.

A different organization are the multiprocessor or parallel processor systems. These computer systems are composed of a set of independent and autonomous processors that are fully or partially interconnected in some way. Multiprocessors can be synchronous or asynchronous where each processor is driven by its own clock. In Flynn's taxonomy these are multiple-instruction-multiple-data or MIMD systems. In MIMD computers the control is distributed, with each processor having its own control unit. Global control units may be used to coordinate the entire system or allow for the design of hierarchical MIMD systems.

Two major subfamilies of parallel processor machines are the shared memory and message passing. In the former organization all processors share the same memory address space, and are connected to a shared physical memory through a high bandwidth bus or a multistage interconnection network. Communication between processors is accomplished through the shared memory. Examples of shared memory systems include the Cray X-MP, Cray 2, ETA-10, Alliant FX/8, and IBM 3090 [5], [9].

In the message passing organization each processor has its own private memory and there is no physical shared memory. Processors communicate asynchronously using message passing mechanisms. Most of message passing systems require a host or some sort of global control unit (which can be one of the processors) that is used to initialize the system and keep global status information. The Intel hypercube, the Caltech cosmic cube, and the N-cube/10 are examples of message passing multiprocessors.

Kuck's taxonomy [19] was a more recent attempt to categorize computer organizations based on their instruction sets. It is more detailed than Flynn's taxonomy, and it will be used through the rest of this paper. According to this taxonomy, the previous systems can be characterized as SEA, MES, or MEA. SEA stands for single-execution-array and includes pipelined and array machines. MES stands for multiple-execution-scalar and is identical to MIMD. Note that MES refers to multiprocessors with scalar processors. A more interesting category is the MEA or multiple-execution-array machines that are multiprocessors with SEA processors. These are generaly more powerful than all previous categories since they provide two levels of parallelism and larger instruction sets. Examples of MEA systems are the Cray X-MP, the Cedar, and the Alliant FX/8.

For this paper MEA organization is implied. We will look at automatic restructuring techniques that vectorize and parallelize serial Fortran programs for MEA computers. Therefore vectorization for each processors, as well as parallelization across processors are of great significance to MEA systems.

3. Parallelism in Programs

In a parallel computer all processors must be used at all times to keep utilization at maximum level. There are two ways one can use to achieve this goal. Either by making the machine available to many users, or by exploiting program parallelism.

In the former (extreme) case, different processors would be executing different user jobs (probably in a timeshared mode). The throughput of the machine may be high, but the individual turnaround time would also be high. In the latter (extreme) case all processors are put to work on a single user program, in which case program execution time is minimized. This can only be done with parallel programs.

One could argue that the primary purpose of designing and building parallel computers is to minimize program execution time (especially for critical applications) as opposed to maximizing throughput [27]. It is well-known that many important questions in science and engineering cannot be answered adequately, simply because their computational needs far exceed the capacity of the most powerful computer systems of today's. It is thus very important to exploit parallelism in such applications. On the other hand, needs for increased throughput may be satisfied by having several different machines, each taking part of the workload. Clearly throughput needs can be satisfied without the use of parallel computers.

For the rest of this paper we assume that parallel programs execute on parallel computers in a batch or dedicated environment, where only one program executes on the machine at any given time.

3.1 Sources and Types of Parallelism

Parallelism can be structured or unstructured. Structured parallelism (e.g., a set of independent identical tasks that operate on different data sets) is more easy to deal with than unstructured parallelism (i.e., different instruction and/or data streams).

At the algorithm design level the user can choose the most appropriate algorithm to carry out a particular computation. This is the first and the most important step in the parallelism specification process. After selecting the most appropriate (parallel) algorithm, the user must select appropriate data structures. For example, in Pascal or in C a set of data items can be stored in a linked list or in a vector. In terms of parallel access, a vector is clearly a more suitable structure since we can address each of its elements independently.

The next step is to code an algorithm using the "most appropriate language". If the language provides parallel constructs (unfortunately very few do) the user may explicitly specify all or some of the parallelism in a program using these language constructs. Alternatively the user may write a program in a serial language and leave the responsibility of parallelism detection to the compiler. A restructuring compiler can also be used to discover more parallelism in a program that has already been coded in a parallel language.

During coding parts of the computation may be organized into subroutines or coroutines. In particular, independent modules of computation can be implemented as independent subroutines. During execution of the program these subroutines can execute in parallel. Parallelism at the subroutine level is *coarse grain*.

At the loop level, parallelism can be specified again by the user or it can be discovered by the compiler through dependence analysis. Parallel loops is one of the most well studied topics in program restructuring. Several different types of parallel loops exist depending on the kind of dependence graph of the loop, and we will define and use most of them in the following sections. Because most Fortran programs spend most of their time inside loops [20], parallelizing loops is one of the most crucial activities during program restructuring and it may result in dramatic speedups [20], [26]. We can classify loop parallelism as *medium grain*.

Parallelism at the basic block level can be medium or *fine grain* depending on the size of the block. Basic blocks can execute concurrently if no interblock dependences exist. In some cases basic blocks with interdependences can also execute in parallel [26]. Fine grain parallelism at the statement or operation level is also important even though the average resulting speedup (at least for numerical programs) is far smaller than at the loop level. Due to the overhead involved, fine grain parallelism is usually exploited inside each processor by utilizing different or multiple functional units, or by using pipelining.

3.2 Expressing Parallelism in Programs

As mentioned above if the programmer has a parallel language at his disposal, then parallelism at all levels can be explicitly specified by the programmer [6], [21]. The higher the level the easier it is to do this. As we go though deeper into each program module, it becomes increasingly cumbersome to specify parallelism even if the language provides all necessary constructs. This is true partly due to the unstructured nature of parallelism at low levels.

An unpleasant fact of life, so far, is that there are no languages available that can be characterized as truly parallel. Even if such languages emerge in the near future (which is inevitable) they will probably be unusually complex for the "casual" programmer. Another problem is the fate of "dusty-decks". In the numerical world the amount of Fortran 66 and Fortran 77 code that is currently in use, translates into billions of dollars investment in human and machine resources. Rewriting these codes is a parallel language would take several years and serious financial commitments.

Restructuring compilers offer a very attractive solution to both of the above problems. If the techniques used to discover parallelism are language independent (which is approximately true) changing a restructuring compiler to generate code for different versions of a programming language (e.g., Fortran) requires little effort. The task of parallelism detection and specification is very complex even for skillful programmers. A restructuring compiler can do a better job than the average programmer, especially with large size programs.

3.3 Utilization of Parallelism

Assuming that parallelism has been specified by the user or has been discovered by the compiler, the next step is to find ways of utilizing this parallelism. That is, find the best mapping of parallel program constructs onto a given parallel architecture. Finding the optimal mapping is a crucial and very complex problem [27].

There are several components in the mapping problem: processor allocation and scheduling, communication, and synchronization. An optimal solution that takes all these components into consideration would result in optimal execution times. However, in general an optimal solution to one of the subproblems implies a suboptimal (and maybe inefficient) solution to the other subproblems. For example, interprocessor communication alone is minimized (in the absolute sense) if the program executes on a single processor. The same is true for synchronization.

A few approaches have been proposed recently that solve some important instances of the mapping problem very efficiently [26], [27]. The complexity of the mapping problem calls for automated solutions through the use of more sophisticated software tools. It is simply not practical to assume that the responsibility lies entirely with the programmer.

3.4 The Role of the Compiler

The complexity of parallel programming increases as the organizations of parallel computers become more and more complex. This complexity makes the need of sophisticated software tools more clear. As it is "natural" for traditional compilers to carry out register allocation and other standard optimizations, it is also more appropriate for serial-to-parallel program restructuring to be the compiler's responsibility. Even though restructuring compilers emerged several years ago, there is still a long way to go before we can consider the problem as adequately solved.

Most current supercomputer vendors supply such compilers for their machines [8], [10], [22], [23]. Several commercial and experimental restructuring compilers are in use including the Vast, the KAP, the PFC, and the Parafrase restructurer which was written at the University of Illinois [4], [8], [12], [16]. All of these compilers focus on program restructuring alone and they ignore other important problems such as scheduling, synchronization, and communication [27]. We believe that these activities can be carried out (at least in part) by the compiler.

Parafrase II, a project that started recently at CSRD, aims towards the development of a powerful parallelizing compiler that it will not only restructure serial programs into a parallel form, but it will also perform communication and synchronization overhead analysis, memory management, and it will help the scheduling of the resulting program for execution on parallel machines. Another goal of this project is to develop a restructurer that will be able to parallelize several different languages. The structure of Parafrase II is shown in Figure 1. Parafrase II is currently being developed to restructure Fortran 77 and C programs for shared memory parallel machines. Long term goals include memory management and automatic analysis of the expected overhead incurred during parallel execution of a given program. The restructurer will also perform the first phase of scheduling called pre-scheduling.

4. Data Dependences

The effectiveness of automatic program restructuring depends on how accurately we can compute data dependences. In general, data dependences give us information about how data are computed in a program and how they are used. This information is then used to parallelize the program automatically. The resulting parallel program should be such that during its parallel execution it computes and uses data in the order implied by the various data and control dependences; this is necessary to guarantee correctness of results. Thus automatic parallelization is done only when the semantics of the original program is preserved (but not necessarily the precision of the results).

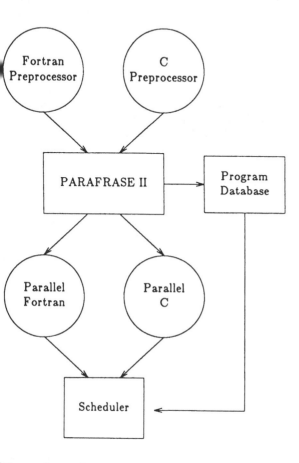

Figure 1. The structure of Paraphrase II.

For each assignment statement in a Fortran program the IN and OUT sets are defined as usual to be the set of input variables (right hand side of the statement) and output variable(s) (left hand side) respectively [1], [2]. The i-th statement in a program is represented with s_i. The order of execution between any two statements s_i and s_j in a program is that which is implied by the serial execution of the program. We say that s_i precedes s_j in the execution order and denote it by $s_i < s_j$, if during serial execution of the program s_i is executed before s_j. Similarly we define $s_i > s_j$, and $s_i = s_j$ when $i = j$. Each statement inside a loop has a number of *instances* equal to the number of iterations of the loop. The above definitions also apply to statement instances.

The first step in a restructuring compiler is to build the *data dependence graph* or DDG or the program. The DDG is a directed graph with nodes representing the statements of the program and arcs corresponding to data dependences. There

are four major types of data dependences in a Fortran program: Flow, anti-dependences, output, and control dependences.

A *flow dependence* defines an order of execution between two statements s_1 and
s_2 iff $OUT(s_1) \cap IN(s_2) \neq \emptyset$ and $s_1 \leq s_2$. In the DDG it is represented by an arc origi-
nating from s_1 which is called the *dependence source* and pointing to s_2 which is
said to be the *dependence sink*. A flow dependence is indicated by $s_1 \delta s_2$. An
anti-dependence from s_1 to s_2 denoted by $s_1 \bar{\delta} s_2$ occurs iff $IN(s_1) \cap OUT(s_2) \neq \emptyset$, and
s_1 precedes s_2 in the serial execution of the program. It is represented in the DDG
by an arc labeled with a dash. An *output dependence* $s_1 \delta^o s_2$ is defined iff
$OUT(s_1) \cap OUT(s_2) \neq \emptyset$, $s_1 \leq s_2$ and in the DDG it is represented by an arc labeled
with the letter "o". Finally we have a *control dependence* between s_1 and s_2
denoted $s_1 \delta^c s_2$ if s_1 is a conditional statement and s_2 is in the scope of s_1. An
example of the four types of dependences is shown in Figure 2.

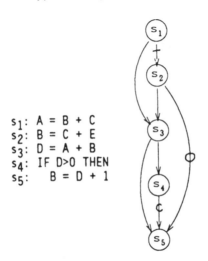

```
s1: A = B + C
s2: B = C + E
s3: D = A + B
s4: IF D>0 THEN
s5:    B = D + 1
```

Figure 2.   An example of data and control dependences.

The example of Figure 2. shows data dependences for scalar variables. The corresponding graph is called the *scalar* dependence graph. When vector elements
are involved (usually inside loops) the computation of data dependences becomes
more complex. We can no longer use the classical use-definition [1], [2]
approach, since a nonempty intersection of the IN and OUT sets does not necessarily result to a true dependence. Subscript analysis is necessary in this case to
determine whether the references in question are to the same vector elements.
Consider for example the following Fortran loop.

```
      DO I = 1, N
s1:      A(I+K) = ...
s2:      ...    =  A(I)
      ENDO
```

where K is a nonnegative integer constant. Here $s_1 \leq s_2$ and $IN(s_1) \cap OUT(s_2) \neq \emptyset$,
but we cannot determine yet whether a flow dependence from s_1 to s_2 exists.
Several factors must be considered in this case. For a dependence to exist we
must have two values of the index I, I_1 and I_2, such that

$$1 \leq l_1 \leq l_2 \leq N \quad \text{and} \quad l_1 + K = l_2.$$

To test this we must know the values of K and N. In most programs the value of K is known at compile-time but this is not always true for loop bounds like N. If $K \leq N$ then a dependence may exist. However if $K > N$ no dependence between the two statements can exist. Frequently loop bounds are not known at compile-time but to be on the safe side they are assumed to be "large". In general, dependences can be computed by solving a Diophantine equation similar to the above. Algorithms for computing data dependences are given in [7], [30].

In the previous example a dependence exists if $l_2 - l_1 = K$. This difference is said to be the *dependence distance* and gives the number of iterations that must be executed between a definition and a use of a particular element of array A (or equivalently, the number of iterations between the iteration containing the source and the iteration containing the sink of the corresponding dependence). The arcs of the data dependence graph involving the statements of a loop are labeled with corresponding dependence distances. Such a graph is said to be the *static* dependence graph of a loop. Similarly the individual dependences are called static. If we unroll the loop we can draw the dependence graph for each individual iteration. Each of these graphs is called an *instance* of the static dependence graph. Instances of the dependence graph are subgraphs of the static DDG. The arcs of these graphs are the instances of the static data dependence arcs. The distance of the various instances of a static dependence arc are not always equal.

For each loop in a program its DDG is built and examined to determine whether that loop can be parallelized. Depending on the structure of the DDG we have different types of loops. First, if the dependence graph forms a cycle that involves all statements of a loop, that loop is serial. A vector loop is one whose individual statements can be rewritten in vector syntax. Another type of parallel loop is the DOALL [13], [24]. A DOALL may be a vector loop or a parallel loop that cannot be vectorized. In any case all iterations of a DOALL can execute in parallel. A more general type of loop is the DOACROSS which allows for partial overlap of successive iterations during execution [11], [24].

5. Automatic Program Restructuring

After determining dependences and building the DDG, the compiler starts the processes of optimizing and restructuring the program from a serial into a parallel form. During the optimization phase several architecture independent optimizations are applied to the source code to make it more efficient, more suitable for restructuring, or both. The restructuring phase which actually transforms the program into a parallel form is also organized into architecture independent and architecture dependent subphases. In this paper we look at program restructuring for SEA and MEA machine organizations.

5.1 Common Optimizations

Some standard and not so standard optimizations are useful independently of the organization of the machine on which we intend to run the program. In this section we discuss some of these optimizations but we cover in no way all of them.

5.1.1 Induction Variable Substitution

Induction variables are often used inside loops to simplify subscript expressions. An *induction variable* is a variable whose values form an arithmetic or geometric progression [1], [2], [25]. These variables are usually expressed as functions of the loop index. Their detection and elimination does not only reduce the number of operations inside the loop, but it may also allow vectorization of the loop that would be impossible otherwise due to dependence cycles. In the following loop

```
DO I = 1, N
   J = 2*I+1
   A(I) = (A(I) + B(J))/2
ENDO
```

the values of J form an arithmetic progression and thus J is an induction variable. The loop can be rewritten as

```
DO I = 1, N
   A(I) = (A(I) + B(2*I+1))/2
ENDO
```

If the values of an induction variable form a progression except at the first or last terms where there might be a discontinuity, then partial loop unrolling at the beginning or at the end of the loop may be necessary to remove the induction variable. This usually happens when the values of such a variable are a permutation of the terms of a progression. For example, in the loop

```
J = N
DO I = 1, N
   A(I) = (B(I) + B(J))/2
   J = I
ENDO
```

the values of J are $N, 1, 2, ..., N - 1$. It is obvious however that J is an arithmetic progression shifted (with wrap around) once to the right. By unrolling the first iteration of the loop we get

```
A(1) = (B(1) + B(N))/2
DO I = 2, N
   A(I) = (B(I) + B(I-1))/2
ENDO
```

5.1.2 Index Recurrences

As is the case with induction variables, some times (less often) some loop-defined index variables are used to index array elements. Their values do not form progressions but they give for example, the sum of the first I terms of a progression, where I is the loop index. An example of a simple index recurrence is shown in the following loop.

```
J = 0
DO I = 1, N
   J = J + I
      A(I) = A(I) + B(J)
ENDO
```

In this case J is the sum $\sum_{k=1}^{I} k$ for each iteration I of the loop. In most cases these variables can be eliminated just like reduction variables. The above loop will become

```
DO I = 1, N
   A(I) = A(I) + B(I*(I+1)/2)
ENDO
```

5.1.3 Loop Unrolling

Loop unrolling can be applied in two forms. In the first, unrolling is done by peeling off one or more iterations at the beginning or at the end of the loop. In the second form unrolling happens inside the loop and results in changing the loop stride [1], [2], [25]. We already showed an application of loop unrolling in the previous section. Unrolling is useful in many other cases. For example, it is often the case where expensive intrinsic function calls such as MIN and MAX are part of DO statements (e.g., DO I=1, MAX(N,M)). If the relationship of N and M is known (which is often the case) the intrinsic function call can be eliminated by peeling off the last iteration of the loop. Consider the loop

```
DO J = 1, N, K
   DO I = J, MIN(J+K, N)
      A(I) = B(I) + C(I)
   ENDO
ENDO
```

where $K \in Z^+$ is the outer loop stride. The intrinsic function call MIN(J+K,N) will be executed $\lceil N/K \rceil$ times. For $N \gg K$ this may introduce a large amount of unnecessary overhead. The compiler can eliminate automatically the intrinsic function call in the above loop by unrolling the last iteration of the outer loop, and produce the following equivalent code:

```
N1 = TRUNC(N/K)
N2 = N1 * K
N3 = N - N2
DO J = 1, N2, K
   DO I = J, J+K
      A(I) = B(I) + C(I)
   ENDO
ENDO
DO I = N3+1, N
   A(I) = B(I) + C(I)
ENDO
```

In its general form loop unrolling replicates the body of the loop inside the loop for K successive values of the loop index, and increments the loop stride by $K - 1$.

5.1.4 Constant Propagation and Expression Evaluation

This is one of the most common optimizations found in almost any conventional compiler [1], [2], [25]. The idea here is to eliminate unnecessary run-time computation that can be easily done by the compiler. This requires the propagation of the value of program constants to all statements in the program that use them. For example, consider the following loop.

```
DO I = 1, N
    PI = 3.14
    PD = 2*PI
    A(I) = PD*R(I)**2
ENDO
```

After constant propagation, constant expression evaluation, and floating of loop invariants, the above loop will be transformed in the more efficient form:

```
PI = 3.14
PD = 6.28
DO I = 1, N
    A(I) = 6.28*R(I)**2
ENDO
```

Another useful application of constant propagation is to replace constants that appear in array index expressions with their actual values. This is a necessary step for computing accurate data dependences.

5.2 Transformations for Vectorizing and Parallelizing Loops

In this section we discuss some known and some new program transformation for discovering and exploiting parallelism in serial programs. There are several different program transformations in use, and each of them is applied to explore a particular type of parallelism, or to take advantage of specific properties of the code and the target machine. Many program transformations are unrelated to each other in the sense that the end-result is the same independently of the order in which they were applied. However, it is more frequently the case where the order in which different program transformations are applied is significant. Thus if we apply the same set of n transformations to the same program $n!$ times, using a different order at each time, we may get $n!$ programs (semantically equivalent but) syntactically different, with varying amounts of parallelism, that could possibly obtain different performance on the same machine. Finding the best order in which transformations should be applied is an open problem and the answer is known for only a few special cases.

The significance of the order of application becomes more evident with more complex machine architectures. For example, the problem is not as significant when we compile for purely vector (SEA) or parallel-scalar (MES) machines, as it is for parallel-vector (MEA) computers. Program restructuring for vector machines is referred to as *vectorization*. Similarly for parallel-scalar machines it is usually called *parallelization*. When the underlying architecture is parallel-vector, then vectorization as well as parallelization are equally important in a good compiler. However in this case we must face many trade-offs between vectorization and parallelization.

In this section we focus on techniques for vectorizing and parallelizing serial programs without emphasizing the trade-offs (except in a few cases). All material that follows assumes that a data dependence graph with all the necessary information has been constructed by the compiler for each specific case and example that is used.

5.2.1 Loop Vectorization and Loop Distribution

When compiling for a vector machine, one of the most important tasks is to generate vector instructions out of loops [8], [13], [25], [30]. [13], 25], [30]. To do this the compiler must check all dependences inside the loop. In the most simple case where dependences do not exist the compiler must *distribute* the loop around each statement of the loop and create a vector statement for each case. Vectorizing the following loop

```
        DO I = 1, N

s1:         A(I) = B(I) + C(I)
s2:         D(I) = B(I) * K

        ENDO
```

would yield the following two vector statements

```
s1:         A(1:N) = B(1:N) + C(1:N)
s2:         D(1:N) = K * B(1:N)
```

where 1:N denotes all the elements 1, 2, ... , N. In the original loop one element of A and one element of D were computed at each iteration. In the latter case however all elements of A are computed before computation of D starts. This is the result of distributing the loop around each statement. In general loop distribution around two statements s_i and s_j (or around two blocks B_i and B_j) is legal if there is no dependence between s_i and s_j (B_i and B_j), or if there are dependences in only one direction. By definition, vectorization is only possible on a statement-by-statement basis. Therefore in a multistatement loop, loop distribution must be applied before vector code can be generated; if distribution is illegal vectorization of multistatement loops is impossible.

In nested loops only the innermost loop can be vectorized. For example, consider the following serial loop.

```
        DO I = 1, N
s1:         A(I+1) = B(I-1) + C(I)
s2:         B(I) = A(I) * K
s3:         C(I) = B(I) - 1
        ENDO
```

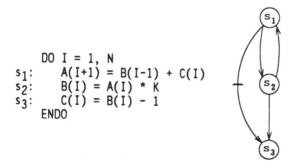

The data dependence graph is shown on the right. A simple traversal of the dependence graph would reveal its strongly connected components (SCC). Those SCCs with single statements (that do not have self-dependences) can be vectorized. The result of vectorizing the above loop would be:

```
DO J = 1, N
    A(I+1) = B(I-1) + C(I)
    B(I) = A(I) * K
ENDO
C(1:N) = B(1:N) - 1
```

5.2.2 Loop Interchange

Loop interchange is another transformation that can be very useful in many cases [25], [30]. It interchanges the (position) nest depth of a pair of loops in the same nest, and can be applied repeatedly to interchange more than two loops in a given nest of loops. As mentioned above, when loops are nested vectorization is possible only for the innermost loop. (Of course, by successive reductions of innermost loops to vector statements, more than one innermost loop can be reduced to vector notation.) Loop interchange can be used in such cases to bring the vectorizable loop to the innermost position. For example, the following loop

```
DO I = 2, N
    DO J = 2, M
    A(I,J) = A(I, J-1) + 1
    ENDO
ENDO
```

is not vectorizable because the innermost loop (a recurrence) must be executed serially. But the outermost loop is parallel. By interchanging the two loops and vectorizing the innermost we get

```
DO J = 1, M
    A(1:N,J) = A(1:N, J-1) + 1
ENDO
```

Loop interchange is not always possible. In general a DOALL loop can be interchanged with any loop nested inside it. The inverse is not always true. Serial loops for example cannot always be interchanged with loops surrounded by them. Interchange is illegal for example in the following loop.

```
DO I = 2, N
    DO J = 1, M
    A(I,J) = A(I-1, J+1) + 1
    ENDO
ENDO
```

In general loop interchange is impossible when there are dependences between any two statements of the loop with "<" and ">" directions [30]. There is no unique way to determine the best interchange. The answer depends on what we want to achieve. The architecture of the target machine for example may dictate which possibility is the best. When we vectorize, interchange should be done so that the loop with the largest number of iterations is brought to the innermost position. This would create vector statements that will operate on long vector operands. For memory-to-memory systems (e.g., CDC Cyber 205) long vectors are particularly important. If on the other hand we compile for a scalar multiprocessor, bringing the largest loop (in terms of number of iterations) in the outermost position is more desirable, since that would allow the parallel loop to use more processors.

5.2.3 Node Splitting and Statement Reordering

Loop vectorization and parallelization is impossible when the statements in the body of the loop are involved in a dependence cycle [1], [2], [30]. Dependence cycles that involve only flow dependences are usually hard to break. There are cases however where dependence cycles can be broken resulting in total or partial parallelization of the corresponding loops. One case where cycle breaking is possible is with dependence cycles that involve flow and anti-dependences. Anti-dependences are "artificial" dependences that can usually be eliminated with simple techniques such as variable renaming.

```
            DO I = 1, N
   s₁:        A(I) = B(I) + C(I)
   s₂:        D(I) = A(I-1) * A(I+1)
            ENDO
                  (a)
```

```
            DO I = 1, N
   s₁:        A(I) = B(I) + C(I)
   s₃:        TEMP(I) = A(I+1)
   s₂:        D(I) = A(I-1) * TEMP(I)
            ENDO
                  (b)
```

```
            DO I = 1, N
              TEMP(I) = A(I+1)
              A(I) = B(I) + C(I)
              D(I) = A(I-1) * TEMP(I)
            ENDO
                  (c)
```

```
            TEMP(1:N) = A(2:N+1)
            A(1:N) = B(1:N) + C(1:N)
            D(1:N) = A(0:N-1) * TEMP(1:N)
                  (d)
```

Figure 3. An example of node splitting and statement reordering

Figure 3 shows an example of breaking a dependence cycle, and its effect on vectorizing the loop. The loop of Figure 3a cannot be vectorized since its statements are involved in a cycle. Node splitting can be employed here to eliminate the cycle by splitting the node (statement) which causes the anti-dependence. As shown in Figure 3b this is done by renaming variable A(I+1) as TEMP(I) and using its new definition in statement s_2. The corresponding dependence graph is shown on the right.

Now that the cycle is broken, loop distribution can be used to distribute the loop around each statement. The loop of Figure 3b can be distributed around s_2 but not around s_1 and s_3, since there are dependences in both directions. Statement reordering can be used here to reorder the statements of the loop as in Figure 3c (reordering is not always legal). The loop of Figure 3c satisfies now the "one direction dependences" rule, and thus it can be distributed around each statement resulting in the three vector statements of Figure 3d.

5.2.4 Code Reordering for Minimizing Communication

In the previous section statement reordering was used to change the direction of dependences in order to make loop distribution possible. Statement (or in general code) reordering can be used in several cases to improve or make possible the application of a certain optimization. For example reordering can be used to improve the effectiveness of DOACROSS loops [11]. Another application of code reordering is in the reduction of interprocessor communication in parallel computers [28].

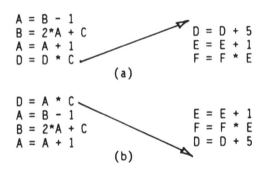

Figure 4. The impact of code reordering in the reduction of communication

The compiler can be used to partition the program into data dependent or independent tasks. Different tasks of a program can execute on different processors, even when there are data dependences across tasks. Is such cases data computed by one or more processors must be communicated to other processors that need them. This communication takes place to satisfy the data dependences and it affects the execution time of the corresponding tasks in the form of processor delays. In certain cases code reordering can be used to reduce or minimize potential communication delays.

Consider for example the two basic blocks of Figure 4a, where each basic block forms a task. The arc in Figure 4a indicates the intertask data dependence. If these two tasks were scheduled to execute simultaneously on two different processors, the processor executing the second task would have to wait until the first task finishes execution. If we assume that each assignment statement takes a unit of time to execute, then the execution time of the two tasks on two processors will be 7 units; the same as if the two tasks were executed on a single processor. If the statements are reordered as in Figure 4b however, the execution time on two processors would be 4 units of time; the second processor will receive the new value of D by the time it will need it. The problem of code reordering for reducing potential interprocessor communication is a complex one and is discussed in more detail in [28].

5.2.5 Loop Blocking

Loop blocking or *strip mining* is a transformation that creates doubly nested loops out of single loops, by organizing the computation in the original loop into chunks of approximately equal size [25]. Loop blocking can be useful in many cases. It is often used to "manage" vector registers, caches, or local memories with small sizes. Many vector computers for example have special vector registers that are used to store vector operands. Loop blocking can be used to partition vector statements into chunks of size K, where K is the length of the vector registers. The following loop

```
DO I = 1, N
    A(I) = B(I) + C(I)
ENDO
```

will become (after blocking)

```
DO J = 1, N, K
    DO I = J, MIN(J+K, N)
    A(I) = B(I) + C(I)
    ENDO
ENDO
```

In the same way blocking can be used to overcome size limitations of caches and local memories.

In parallel-vector machines (MEA) operations with long vector operands can be partitioned into shorter vectors, assigning each of the short vectors to a different processor. In this case loop blocking will introduce a parallel loop as in the following example. Consider the vector statement.

```
A(1:N) = B(1:N) * C(1:N)
```

In a system with P-processors (and if $N \gg P$), this vector operation can be speeded up further by blocking it as follows.

```
K = TRUNC(N/P)
DOALL I = 1, P
    A((I-1)K+1:IK)  = B((I-1)K+1:IK) * C((I-1)K+1:IK)
ENDO
A(PK+1:N) = B(PK+1:N) * C(PK+1:N)
```

Notice that loop unrolling was implicitly used to eliminate the use of the intrinsic function MIN in the DOALL statement.

5.2.6 Cycle Shrinking

Here we discuss a new transformation that can "squeeze" parallelism out of some types of serial loops. Loops whose statements are involved in a dependence cycle are considered to be serial by most of the existing commercial and experimental compilers. This is particularly true if all dependences in the cycle are flow dependences where node splitting is unable to help. Depending however on the distance of each dependence we may parallelize in part some of these loops. *Cycle shrinking* can achieve this as explained below.

```
DO I = 3, N
    A(I) = B(I-2) - 1
    B(I) = A(I-3) * K
ENDO
        (a)

DO J = 3, N, 2
    DOALL I = J, J+1
        A(I) = B(I-2) - 1
        B(I) = A(I-3) * K
    ENDOALL
ENDO
        (b)
```

Figure 5.   An example of the application of cycle shrinking.

It is clear that iterations of a loop that are free of dependence sinks can execute in parallel. Cycle shrinking partitions loop iterations into groups that are free of dependence sinks, assuming that all groups with earlier iterations have completed execution before the current group is scheduled. By checking the dependence distances at compile-time we can determine the size of these iteration groups and then partition the loop (by blocking it) in the appropriate way. In [28] it is shown that the best result can be achieved if the size of the iteration groups is equal to the minimum dependence distance in the dependence cycle of a given loop. A similar scheme is described in [24] based on a greatest common divisor test; but our approach is always better than that of [24].

Figure 5b shows an example of the application of the cycle shrinking transformation in the loop of Figure 5a. The two flow dependences in the loop of Figure 5a have distances of 2 and 3 respectively. Thus the dependence sink-free iteration groups are of size 2. This results in the formation of the innermost DOALL loop that executes in parallel two iterations at a time. The outermost loop now has a stride of 2. In general, the stride of the outermost loop is equal to the cycle reduction factor. The resulting loop in Figure 5b can be executed twice as fast as the original serial loop. A more detailed presentation of cycle shrinking is given in [28].

5.2.7 Loop Coalescing

Most of the existing parallel processor systems are capable of executing in parallel only single loops. However, in real Fortran programs we often observe several nested loops that are all parallel. *Loop coalescing* can transform multiply nested parallel loops into singly nested parallel loops for execution on parallel machines that support only single parallel loops (e.g., Cray X-MP, Alliant FX/8). Coalescing of nested loops is achieved through a universal index mapping that linearizes multidimensional iteration spaces [27], [28], [29].

```
DOALL 1 J=1 , N
   DOALL 2 K=1 , N

      A(J,K) = ...

2      ENDOALL
1  ENDOALL
```

(a)

```
DOALL 1 I=1,N²

  A(⌈I/N⌉, I - N ⌊(I-1)/N⌋) = ...

1  ENDOALL
```

(b)

Figure 6.   Loop coalescing in two dimensions.

Another application of loop coalescing is in static processor allocation [29]. In systems that perform processor allocation statically through the compiler or the operating system, deciding what is the best allocation of processors in a multi-nested parallel loop is a complex and expensive process. If multinested loops are coalesced into single loops, the static processor allocation problem becomes straightforward.

Figure 6 shows an example of the application of loop coalescing to a doubly nested parallel loop. In this case the original loop of Figure 6a is perfectly nested even though coalescing can be applied to non-perfectly nested loops. The transformation maps the individual indices of array references in the original loop, into expressions that involve a single common index, i.e., that of the coalesced loop. Figure 6b shows how the indices J and K of the original loop can be expressed as functions of I, the index of the transformed loop. The transformation is applicable to any polynomial index expressions. A more thorough description of loop coalescing is given in [29].

6. Run-Time Dependence Testing

So far data dependence testing has been carried out exclusively by the compiler. Some of the existing methods can detect all real dependences inside loops [7]. There are many cases however where dependence testing by the compiler is not as accurate and it is usually conservative. Loops whose static dependence graph forms a cycle are considered to be serial by most of the existing compilers. In [28] a different approach for computing data dependences is described, which performs run-time checks to detect dependence sinks that have not been satisfied.

This approach which is briefly described in this section, achieves better results in parallelizing certain loops for which compile-time schemes fail.

```
DO I = 1, N
    A(2I-1) = B(I-1) + 1
    B(2I+1) = A(I+1) * C(I)
ENDO
    (a)

DO I = 1, N
    IF (s₁(I+1)<>0) AND (s₁(I+1)<I) WAIT ON V₁(I+1)
    IF (s₂(I-1)<>0) AND (s₂(I-1)<I) WAIT ON V₂(I-1)
    A(2I-1) = B(I-1) + 1
    B(2I+1) = A(I+1) * C(I)
    CLEAR V₁(2I-1)
    CLEAR V₂(2I+1)
ENDO
    (b)
```

Figure 7.   The loop af (a) is transformed to that of (b) after RDC transformation.

Run-time dependence checking is done in two phases. In the first phase the compiler computes potential dependence sources and stores them in an auxiliary vector S. At the same time the compiler creates a bit-vector V which is used at run-time to synchronize the execution of loop iterations wherever appropriate. In the second (run-time) phase, all memory references to array elements that may constitute a dependence sink are checked against the potential dependence sources. If a match is found the synchronization vector is used to hold the execution of that statement until its corresponding dependence source is computed. If a match does not occur the corresponding iteration is immediately scheduled for execution. This allows the concurrent execution of loops that are considered as serial by compilers that use compile-time dependence checking only.

An example of run-time dependence checking is shown in Figure 7 The static dependence graph of the loop of Figure 7a involves a cycle and the loop is therefore considered to be serial. If the *run-time dependence checking* or RDC transformation is applied the following actions will take place. First the compiler will create and compute the source vectors s_1 and s_2 for the potential dependence sources in the original loop (in this case A(2I-1) and B(2I + 1)). With each source vector we associate a bit-vector (V_1 and V_2 respectively). The compiler then executes the following loop:

```
DOALL I=1, N
    S₁(2I-1) = I
    V₁(2I-1) = 1
    S₂(2I+1) = I
    V₂(2I+1) = 1
ENDO
```

After the source and the synchronization vectors are computed, the compiler transforms the original loop in the loop of Figure 7b, where the WAIT statement implies a synchronization operation that blocks the execution of a particular iteration until the corresponding bit position in the V vector is cleared. If the corresponding synchronization bit is cleared execution proceeds in the normal way. The CLEAR statement in the loop of Figure 7b resets the corresponding bit of vector V to zero. Run-time dependence checking [28] blocks only the iterations that are involved in an unsatisfied dependence until the corresponding dependence source is computed. All other iterations of the loop are executed in an asynchronous concurrent mode. This is not possible when the conventional methods for computing data dependences are used.

7. Conclusions

The restructuring compilers of the future will undoubtly need to possess many properties that we do not find in today's most sophisticated parallelizing compilers. So far, the attention has only focused on optimization and restructuring techniques. However, the complexity of the new parallel machines requires the compiler to perform many additional functions than just restructuring. These functions include memory management in a memory hierarchy environment. Good memory management requires knowledge of the program which the operating system is unable to know. Scheduling is also a candidate for the compilers of the near future. Memory management, minimization of interprocessor communication, synchronization and various other types of overhead are important issues that could be tackled by the compiler. Another important aspect of the near future compilers for parallel machines is ease of use and interaction with the user. There are many cases where the user's assistance (in the form of assertions for example) is necessary for parallelizing a program and exploiting the resulting parallelism.

It is our belief that in the next few years we will see a transfer of many run-time activities (that are now considered the operating system's responsibility), to the compiler. This will become necessary as performance becomes more of a critical factor. Any activity involving the operating system is known to involve a large overhead. This overhead cannot be tolerated above a certain point. Also, in time-sharing, the system's knowledge of specific program characteristics is not necessary to achieve high throughput. In parallel processor environments however, knowledge of program characteristics is necessary for minimizing program turnaround time. Thus the shift of operating system functions to the compiler will be a logical consequence. Compilers will become highly interactive and far more complex than modern restructurers, while the software layer between the user and the hardware called the operating system will become thinner, at least in high performance computer systems.

8. References

[1]   A.V. Aho, R. Sethi, and J.D. Ullman,
        Compilers: Principles, Techniques, and Tools,
        Addison-Wesley, Reading, Massachusetts, 1986.

[2] F.E. Allen and J. Cocke,
"A Catalogue of Optimizing Transformations,"
Design and Optimization of Compilers, R. Rustin, Ed. Prentice-Hall, Englewood Cliffs, N.J., 1972, pp. 1-30.

[3] F.E. Allen, J.L. Carter, J. Fabri, J. Ferrante, W.H. Harrison, P.G. Loewner, and L.H. Trevillyan,
"The Experimental Compiling System,"
IBM J. Res. Dev. 24, 6, November 1980, pp. 695-715.

[4] J.R. Allen and K. Kennedy,
"PFC: A Program to Convert Fortran to Parallel Form,"
Techn. Rept. MASC-TR82-6, Rice University, Houston, Texas, March 1982.

[5] Alliant Computer Systems Corp.,
"FX/Series Architecture Manual," Acton, Massachusetts, 1985

[6] American National Standards Institute,
American National Standard for Information Systems. Programming Language Fortran S8 (X3.9-198x). Revision of X3.9-1978, Draft S8, Version 99, ANSI, New York, April 1986.

[7] U. Banerjee,
"Speedup of Ordinary Programs,"
Ph.D. Thesis, University of Illinois at Urbana-Champaign, DCS Report No. UIUCDCS-R-79-989, October 1979.

[8] B. Brode,
"Precompilation of Fortran Programs to Facilitate Array Processing,"
Computer 14, 9, September 1981, pp. 46-51.

[9] S. Chen,
"Large-scale and High-speed Multiprocessor System for Scientific Applications - Cray-X-MP-2 Series,"
Proc. of NATO Advanced Research Workshop on High Speed Computing, Kawalik (Editor), pp. 59-67, June 1983.

[10] "Multitasking User Guide,"
Cray Computer Systems Technical Note, SN-0222, January, 1985.

[11] R.G. Cytron,
"Doacross: Beyond Vectorization for Multiprocessors (Extended Abstract),"
Proceedings of the 1986 International Conference on Parallel Processing, St. Charles, IL, pp. 836-844, August, 1986.

[12] J. Davies, C. Huson, T. Macke, B. Leasure, and M. Wolfe,
"The KAP/205: An Advanced Source-to-Source Vectorizer for the Cyber 205 Supercomputer,"
Proceedings of the 1986 International Conference on Parallel Processing, St. Charles, Illinois, August, 1986.

[13] J. R. Beckman Davies,
"Parallel Loop Constructs for Multiprocessors,"
M.S. Thesis, University of Illinois at Urbana-Champaign, DCS Report No. UIUCDCS-R-81-1070, May, 1981.

[14] J. J. Dongarra,
"Comparison of the CRAY X-MP-4, Fujitsu VP-200, and Hitachi S-810/20: An Argonne Perspective,"
Argonne National Laboratory, ANL-85-19, October, 1985.

[15]   K. Kennedy,
       "Automatic Vectorization of Fortran Programs to Vector Form,"
       Technical Report, Rice University, Houston, TX, October, 1980.

[16]   D.J. Kuck, R.H. Kuhn, B. Leasure, and M. Wolfe,
       "The Structure of an Advanced Vectorizer for Pipelined Processors,"
       Fourth International Computer Software and Applications Conference, October,
       1980.

[17]   D.J. Kuck, R. Kuhn, D. Padua, B. Leasure, and M. Wolfe,
       "Dependence Graphs and Compiler Optimizations,"
       *Proceedings of the 8-th ACM Symposium on Principles of Programming Lan-
       guages*, pp. 207-218, January 1981.

[18]   D. J. Kuck, E. S. Davidson, D. H. Lawrie, and A.H. Sameh,
       "Parallel Supercomputing Today and the Cedar Approach,"
       Science 231, 4740 February 28, 1986, pp. 967-974.

[19]   D.J. Kuck,
       The Structure of Computers and Computations, Volume 1, John Wiley and Sons,
       New York, 1978.

[20]   D.J. Kuck et. al.,
       "The Effects of Program Restructuring, Algorithm Change and Architecture Choice
       on Program Performance,"
       International Conference on Parallel Processing, August, 1984.

[21]   P. Mehrotra and J. Van Rosendale,
       "The Blaze Language: A Parallel Language for Scientific Programming,"
       Rep. 85-29, Institute for Computer Applications in Science and Engineering, NASA
       Langley Research Center, Hampton, Va., May 1985.

[22]   K. Miura and K. Uchida,
       "Facom Vector Processor VP-100/VP-200,"
       High Speed Computation, NATO ASI Series, Vol. F7, J.S. Kowalik Ed., Springer-
       Verlag, New York, 1984.

[23]   S. Nagashima, Y. Inagami, T. Odaka, and S. Kawabe,
       "Design Consideration for a High-Speed Vector Processor: The Hitachi S-810,"
       *Proceedings of the IEEE International Conference on Computer Design: VLSI in
       Computers*, ICCD 84, IEEE Press, New York, 1984.

[24]   D.A. Padua Haiek,
       "Multiprocessors: Discussions of Some Theoretical and Practical Problems,"
       Ph.D. Thesis, University of Illinois at Urbana-Champaign, DCS Report No.
       UIUCDCS-R-79-990, November 1979.

[25]   D.A. Padua, and M. Wolfe,
       "Advanced Compiler Optimizations for Supercomputers,"
       Communications of the ACM, Vol. 29, No. 12, pp. 1184-1201, December 1986.

[26]   C. D. Polychronopoulos and U. Banerjee,
       "Processor Allocation for Horizontal and Vertical Parallelism and Related Speedup
       Bounds,"
       IEEE Transactions on Computers, **Special Issue on Parallel and Distributed Com-
       puting**, Vol. C-36, No. 4 April, 1987.

[27]   C. D. Polychronopoulos and D. J. Kuck,
       "Guided Self-Scheduling: A Practical Scheduling Scheme for Parallel Supercom-
       puters,"
       to appear *IEEE Transactions on Computers*, **Special Issue on Supercomputing**,
       December, 1987.

[28]   C. D. Polychronopoulos,
        "On Advanced Compiler Optimizations for Parallel Computers,"
        Proceedings of the International Conference on Supercomputing, (Athens, Greece,
        June 8-12, 1987), E. Houstis, T. Papatheodorou, and C.D. Polychronopoulos Ed.,
        Springer-Verlag, New York, 1987.

[29]   C. D. Polychronopoulos,
        "Loop Coalescing: A Compiler Transformation for Parallel Machines,"
        Proceedings of the 1987 International Conference on Parallel Processing, St.
        Charles, Illinois, August, 1987.

[30]   M. J. Wolfe,
        "Optimizing Supercompilers for Supercomputers,"
        Ph.D. Thesis, University of Illinois at Urbana-Champaign, DCS Report No.
        UIUCCDCS-R-82-1105, 1982.

Part IV - Algorithms and Applications

Locating Parallel Numerical Tasks in the Solution of Viscous Fluid Flow

Garry Rodrigue
Lawrence Livermore National Laboratory,
University of California, Davis

A. Louise Perkins
University of California, Davis

Summary

We present a method to solve the heat equation that couples mesh refinement with explicit time steps greater than the Courant condition limit. The method is implemented in parallel and executes efficiently.

1. Introduction

On parallel computers, it is important to have well-defined numerical tasks that can be executed simultaneously. For partial differential equations, the idea of domain decomposition has been shown many times to be an effective means for defining such tasks [3], [4]. Basically, a partial differential equation

$$F(x,u,t,D_t u,D_x u, D_x^2 u, \dots) = 0$$

is defined where $x \in \Omega \subset \mathbb{R}^n$, $0 < t \le \tau$, and $u(x,t) \in \mathbb{R}^m$ satisfy given boundary and initial conditions. Then at a given computational time t^*, the domain Ω is expressed as the union of subdomains (not necessarily disjoint)

$$\Omega = \bigcup_{j=1}^{k(t^*)} \Omega_j (t^*).$$

On each subdomain $\Omega_j (t^*)$, a partial differential equation

$$F_j (x,u_j, t, D_t u_j, D_x u_j, D_x^2 u_j, \dots) = 0$$

is defined where $x \in \Omega_j(t^*) \subset \mathbb{R}^n$, $t^* < t \le t^* + \Delta t_j$, and $u_j (x,t) \in \mathbb{R}^m$ satisfies given boundary and initial conditions on $\Omega_j (t^*)$. The connection between domain decomposition and parallel computers is that each processor will have the task of solving a problem $F_j = 0$ and conveying numerical information about its computed solution to neighboring processors.

How one defines the particular decomposition of the domain is a difficult problem and depends very much on the properties of the solution. In the following sections we outline a method for locating the subdomains when the partial differential

This work was performed under the auspices of the U.S. Department of Energy by the Lawrence Livermore Laboratory under Contract W-7405-Eng-48.

equation describes viscous fluid motion. The idea also has application to any partial differential equation governing conservation laws where diffusivity is allowed.

2. The Equations

In this paper we demonstrate a numerical technique for solving, on a parallel computer, the two dimensional heat equation over a small time interval Δt with an initial condition that possesses a sharp gradient. This problem is motivated from the study of the numerical solution of viscous fluid equations. In this situation, the two-dimensional Navier-Stokes equation for momentum is

$$(2.1) \qquad \frac{\partial V}{\partial t} + V \cdot \nabla V = \mu \nabla^2 V + g(x,y)$$

where μ is the coefficient of viscosity and V is the velocity vector. Initial conditions for these equations are frequently discontinuous and, in the case when $\mu = 0$, these discontinuities are propagated with time. In the case when $\mu > 0$, the solution develops layers around the discontinuity. However, unlike the inviscid case, these layers can propagate or dissipate with time.

In an inertial Cartesian coordinate system, the two components $a(u,v)$, $b(u,v)$ of the term $V \cdot \nabla V$ in (2.1) (here, $V = [u,v]'$) are given by

$$a(u,v) = u\frac{\partial u}{\partial x} + v\frac{\partial u}{\partial y},$$

$$b(u,v) = u\frac{\partial v}{\partial x} + v\frac{\partial v}{\partial y}.$$

The above equations are based on the Eulerian approach for the description of the continuum motion: the characteristic properties of the medium (density, velocity, etc.) are considered as functions of time and space in the frame of reference. An alternative description is provided by the Lagrangian formulation in which the dependent variables are the characteristic properties of material particles that are followed in their motion: these properties are thus functions of time and parameters used to identify the particles, such as the particle coordinates at some fixed initial time. In the latter situation, the equations of motion describe the properities and position of this particle as the fluid changes with time.

If (x,y) is the coordinates of a particle at the initial t_0 and

$$(2.2) \qquad R(x,y;t) = \begin{bmatrix} \xi \\ \eta \end{bmatrix} = \begin{bmatrix} \xi(x,y;t) \\ \eta(x,y;t) \end{bmatrix}$$

is the coordinates at a later time, then for fixed time t, R defines a change of coordinates. The Jacobian of R is

$$J(x,y) = \begin{bmatrix} \partial\xi/\partial x & \partial\xi/\partial y \\ \partial\eta/\partial x & \partial\eta/\partial y \end{bmatrix}$$

and we assume $j(x,y) = determinant\,[J(x,y)] > 0$

It then follows (cf. [3]) that

$$\nabla^2 u = j^{-1}(x,y)\left\{\frac{\partial}{\partial x}\,a(x,y)\frac{\partial u}{\partial x} + \frac{\partial}{\partial y}\,b(x,y)\frac{\partial u}{\partial y}\right.$$
$$\left. - \frac{\partial}{\partial y}\,c(x,y)\frac{\partial u}{\partial x} - \frac{\partial}{\partial x}\,c(x,y)\frac{\partial u}{\partial y}\right\}$$

(2.3)

where

$$a(x,y) = j^{-1}(x,y)\left[(\partial\eta/\partial y)^2 + (\partial\xi/\partial y)^2\right]$$
$$b(x,y) = j^{-1}(x,y)\left[(\partial\eta/\partial x)^2 + (\partial\xi/\partial x)^2\right]$$
$$c(x,y) = j^{-1}(x,y)\left[\frac{\partial\eta}{\partial x}\frac{\partial\eta}{\partial y} + \frac{\partial\xi}{\partial x}\frac{\partial\xi}{\partial y}\right]$$

A similar expression holds for $\nabla^2 v$. Since

$$V = \frac{dR}{dt}$$

and

$$\frac{dV}{dt} = \frac{\partial V}{\partial t} + \nabla V \cdot \frac{dR}{dt}$$

we see from (2.3) that equation (2.1) becomes

(2.4)
$$\frac{dV}{dt} = \mu j^{-1}(x,y)\nabla \cdot J(x,y)\nabla V + g$$

(2.5)
$$\frac{dR}{dt} = V$$

where R is given by (2.2). Thus, (2.1) is transformed to the two equations above. Thus, equation (2.1) is equivalent to (2.4)-(2.5) in which the diffusion equation (2.4) is an important part.

3. The Heat Equation

Because of the importance of the diffusion equation in the solution of (2.4) - (2.5), we investigate the model heat equation,

(3.1)
$$V_t = \nabla^2 V,$$

with initial conditions that are constant on concentric disks about the origin. That is,

$$V(x,y,0) = V_i$$

where V_i is the constant temperature of the ith disk of radius r_i. For simplicity we examine the one disk case: $r_i = r$, and the problem domain is the unit square, see Figure 1.

Subdomain activity is located with an explicit time step on a coarse mesh that exceeds the Courant condition limit. This introduces high frequency errors into our solution that are manifested as oscillations. This explicit solution is checked for

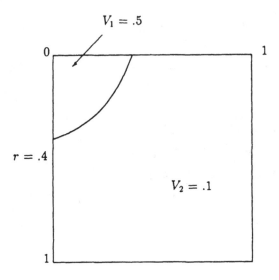

$V_1 = .5$

0   1

$r = .4$

$V_2 = .1$

1

Figure 1.   (Initial Conditions)

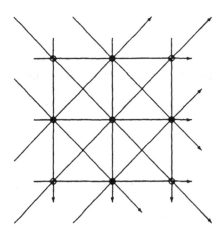

Figure 2.   Adjacent points compared

rapid changes. We use the gradient decomposition method, [2], to locate oscil-
lation by comparing adjacent points against a threshold slope. This was done in
four one dimensional passes so that each neighboring grid point would be considered and is illustrated in Figure 2. If they exceed the threshold they are
flagged. All such areas are identified, clustered, and then the clustered areas are
uniformly refined to allow better resolution of a discontinuity. This local uniform

mesh refinement was expounded in [1]. We use it here because of its programming simplicity and efficiency.

An implicit step is then taken within the finer grids on the subdomains. Currently, a Gauß-Seidel splitting is used to "solve" the resulting implicit matrix problem. Once each refined grid's matrix equation has converged to a solution, any adjoining and overlapping grids undergo an outer iteration to insure they agree on shared grid points. If they do not, a Schwarz alternating update procedure is initiated, and the individual grids are then scheduled in parallel for another inner iteration.

The heat equation is differenced centrally in space and a forward Euler time differencing is used to locate the subdomains. On the subdomains, the implicit backward Euler method is used. The Schwarz Alternating Method, [3], is used to solve the implicit equations for overlapping subdomains. Neumann Boundary Conditions are imposed on (3.1) and we use a mesh, $\Delta x = \Delta y = .1; \Delta t = .01$.

Using a straight forward clustering algorithm that selects axis-aligned rectangular grids from the mesh points indicating oscillations, the program generated three subdomains along the concentric front. These areas are forced to overlap because they are physically adjacent and must provide each other with valid boundary conditions for the Schwarz process. This is illustrated in Figure 3. Dirichlet boundaries for the subdomains are obtained by linear interpolation of coarse grid data (except if a subdomain boundary portion abuts the problem boundary in which case the boundary condition is Neumann.)

```
111111111111122222222111111111111       1122222222222222222221111111111111111
1            2        2          1       1 2       2        2  2              1
1      22222222       2          1       1 2       2        2  2              1
1      2      2       2          1       1 2       2        2  2              1
1      2      2       2          1       2222222222          2  2              1
1      2      2       2          1       2 2       2        2  2              1
22222222      2       2          1       2 2       2        2  2              1
2      2      22222222           1       2 2       2        2  2              1
2      2      2                  1       2 2       2        22222222222       1
2      22222222                  1       2 2       22222222222              1
22222222                         1       2222222222          2              1
1                                1       1 222222222222222222               1
1                                1       1                                   1
1                                1       1                                   1
1                                1       1                                   1
1                                1       1                                   1
1                                1       1                                   1
1                                1       1                                   1
1                                1       1                                   1
1                                1       1                                   1
1                                1       1                                   1
1                                1       1                                   1
1                                1       1                                   1
111111111111111111111111111111111       11111111111111111111111111111111111111
```

Figure 3.   Clustering and overlapping subgrids

At the end of the time step the refined grid solutions are superimposed onto the coarse (parent) grid. We then iterate the coarse grid to convergence also. Only a few iterations are needed on the coarse mesh to converge. This compares to the hundreds of iterations needed to converge the coarse mesh without the mesh refinement.

The results shown were computed on a Cray X-MP.

4. Programming Issues

Adaptive domains vary with the evolving solution. This introduces some practical programming issues to address. Following [1] we use logical trees to relate the many subdomains. To optimize tree traversal for this problem and algorithm, we introduce an inverted tree to physically represent the logical tree. We also dupli- cate variables used by all grids to avoid having to lock shared variables. We found an optimal scheduling algorithm for the subgrids with logical tree relationships. Lastly we managed our memory heap in two dimensions to enhance program readability.

4.1 Inverted Tree

Tree structures are frequently used to monitor mesh refinement in many areas of computer science. To make local uniform mesh refinement computationally attractive, the overhead of managing the mesh must be kept as small as possible. Borrowing from the field of Data Base Management, we extract our tree informa- tion into an index. This is done to better utilize current available hardware and for code readability.

A tree has a parent or root node pointing to all of its children. Its nodes may be structured like Table 1. There are many ways to implement these logical con- nections physically. We designed an inverted tree to best model our access requirements. Its structure is given in Figure 4.

Loc.	Num. Children	Child Pointer	Child Pointer

Table 1. Popular Node Structure

The inverted tree is functionally equivalent to the logical tree when the order of the siblings is not important. This is easily shown by constructing either from the oth- er. The index allows tree traversal to occur in a local memory section, perhaps a cache. At birth a parent knows where the child is, so it does not need to get its address. At birth, the parent gives the child the parent's address and a task to complete. When the child node has completed its task, it informs its parent. Our communication is from child to parent. Thus the physical structure, is an inverted tree. The node record structure is static and no variable length linked lists are needed.

The inverted tree is very efficient for leaf to root traversals, but not as efficient in the other direction. For the local uniform mesh refinement running in parallel, the

Index

Logical Tree

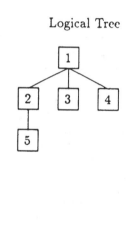

Node Number	Ptr. to Parent
1	0
2	1
3	1
4	1
5	0

Figure 4.   Inverted Tree Structure

majority of communication flows this way. Thus, this structure makes more efficient use of the current hardware available and better reflects well trod software communication paths.

A side effect is readability! In tree format, attributes are buried within pointers to pointers. The code accessing these attributes does so indirectly. For example,

Tree(parent(child(attribute))) .

The use of the index alleviates this problem by replacing the linked list access path by a node number. Thus we are free to build attribute structures directly and access them as follows,

attribute(current node) .

4.2 Duplicate variables vs Shared Variables

Each grid exists in parallel with other grids, but not with itself. That is, the inverted tree node position is assigned a grid for its lifetime. The same is true for all variables used to control subroutine execution such as do loop counters and stack pointers, as well as stacks. We make duplicates of these variables and records so that grid execution can proceed in parallel without shared variable lockage. This is functionally equivalent to what a reentrant code generating compiler would do for us if we could find one.

4.3 The Scheduling Algorithm

To introduce parallelism into the solution of the independently advancing grids, we must give up close parent supervision and free the children to do their own tasks within the parameters established by their parent. The parent creates each child grid and knows the maximum time steps the child will advance before reaching the synchronization barrier. The parent need not synchronize with the child at each shared time step; the two proceed completely independently. They wait for each other at the synchronization barrier. Only overlapped siblings need to synchronize their time advancement.There seems to be two ways of maximizing parallelism for a given problem. The first, is to maximize the number of independent tasks. The second, to minimize the number of barriers. To minimize the number of barriers, a rule of thumb suffices: never wait for something until you need it to proceed. This is more frequently referred to as Data Dependence. If a program synchronizes only at data dependence points it will maximize its parallelism. This doesn't say that a better division of subtasks wouldn't work even better.

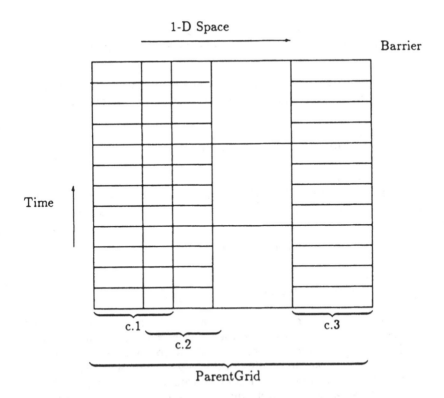

Figure 5.   1-D Scheduling Constraints

To illustrate our scheduling algorithm built upon the above rule of thumb, consider the chart that represents grids advancing in time in Figure 5. If a child can advance to the synchronization barrier undisturbed, as can child 3, it does. No synchronization occurs until the child arrives in time at the barrier. The parent is free to advance to the barrier. However, child 1 and child 2 are "siamese twins" and must proceed with barriers at each time step. There they will undergo an outer iteration and possibly be forced to return for another inner iteration step.

5. Timing Studies

The grid configuration is static from Barrier to Barrier. Consider this configuration as a graph with two types of edges, directed and undirected. That is,

$$G = (V, E_D, E_u),$$

where V is the grids, E_D, the directed edges reflecting the child to parent relation, and E_u, the undirected edges representing overlapping grids at the same level.

To schedule grid advancement in parallel we consider only the subgraph

$$G_u = (V, E_u).$$

This graph can consist of several disjoint subgraphs. For example, if out of 6 grids, 3 and 4, and 5 and 6 overlap, then there are four disjoint subgraphs: {1}, {2}, {3,4}, {5,6}. All of these subgraphs can be advanced in parallel to the barrier.

Within each of the disjoint subgraphs each grid undergoes a number[1]) of inner iterations. These can be done in parallel. Then overlapping nodes must undergo an outer (Schwarz) iteration. These outer iterations can also proceed in parallel.

Let t_i^m be the CPU time required to advance grid i in one inner iteration to time step m. Let the time step count be reset to one after each barrier and K be the number of coarse time steps, and let the temporal refinement ratio be ℓ. Let the coarse grid be grid number 1, and only allow one level of refinement[2]). Further let $t_{S_{ij}}$ be the time required for a Schwarz iteration between grids i and j of which there are N overlapping sections, and S_m be the number of Schwarz iterations needed at time step m for G refined grids. Then the parallel advancement time is computed as follows,

$$t_p = \sum_{m-1}^{K} t_1^m + \sum_{m-1}^{K \cdot \ell} \left(\sum_{m-1}^{S_m} \max_i(t_i^m) + \sum^{S_m} \max_{ij}(t_{S_{ij}}) \right).$$

The serial time required is given by

$$t_s = \sum_{m-1}^{K} t_1^m + \sum_{m-1}^{K \cdot \ell} \left(\sum^{S_m} (\sum_{i-1}^{G} t_i^m) + \sum^{S_m} (\sum_{ij}^{N} t_{S_{ij}}^m) \right).$$

We calculate speed up as a ratio:

$$speedup = \frac{t_s}{t_p}.$$

---

[1] To convergence or a ceiling limit.
[2] This is not necessary but it simplifies the presentation.

Thus a speed up of unity indicates equivalent time required for the serial and parallel execution.

The parallelism timings are simulated. We collected the above required timing values using the system utility TIMEUSED. For the timing results of interest we ran the problem with and without mesh refinement. Thus we have three test cases: coarse mesh in serial, coarse mesh with refined meshes in serial, and the coarse with refined meshes in parallel. We also counted the total number of grid iterations for the coarse and refined meshes. This was to give us an indication of where we were spending our time and allow us to predict how the method would scale up to a larger problem.

Lastly we kept track of all costs associated with mesh refinement. This included deciding which areas need to be refined, refining them, generating new grids for the refined areas as well as disposing of them when we are done. This gave us an indication of how much time was spent in managing our grid structure.

	Coarse G.S.	Refined G.S.
Total Comp. Time		
Serial	.68351	.96733
Parallel	N/A	.44875
"Speed Up"	1	2.15
Grid Overhead	N/A	.05272
Parent Iterations	194	10
Child Iterations	N/A	400

Table 2.   Timing Results

Table 2 presents our timing results as well as an analysis of where we spent our time. We see that the number of iterations for the global domain has been significantly reduced by the addition of mesh refinement. The number of child iterations, however, is quite large. This is in part due to the refinement in time as well as space. The child grid's time step is 1/4 of the parent's. Also, as shown in Figure 3 on page 137, about 1/4 of the coarse mesh has been refined. We believe this percentage will be significantly reduced when we use the finer coarse mesh that the problem requires. If so, this would tip the scales much more heavily for the refined approach. As it is, the parallel version of the refined variation runs about 1/3 faster than the coarse grid advancement without refinement even when we include the time to manage these grids. We are encouraged by these results.

6. References

[1]   Berger, Marsha J. and Oliger, Joseph,
       Adaptive Mesh Refinement For Hyperbolic Partial Differential Equations,
       Journal of Computational PHysics 53, 484-512 (1984).

[2]   Bolstad, John H.,
        An Adaptive Finite Difference Method for Hyperbolic Systems in one space
        dimension,
        Lawrence Berkeley Lab LBL-13287-rev., (1982).

[3]   Rodrigue, G., Kang, L., Lin, G., and Wu, Z.,
        The Generalized Schwarz Alternating Principle,
        to be published.

[4]   Rodrigue, G.,
        Inner/Outer Iterations and Numerical Schwarz Algorithms,
        Journal of Parallel Computing Vol. 2, pp 205-218, (1985).

[5]   Rodrigue, G.,
        An Implicit Numerical Solution of the Two-Dimensional Diffusion Equation and
        Vectorization Experiments,
        Parallel Computatins, Academic Press, 101-128, (1982).

On the Multigrid Acceleration Approach in Computational Fluid Dynamics

Karl Solchenbach[1][2]

Ulrich Trottenberg[1][2][3]

Abstract

In this short note, two multigrid approaches for the treatment of computational fluid dynamics problems are distinguished: the "optimal approach", where the specific model is to be treated entirely by multigrid and all multigrid components are to be defined optimally tailored - versus the "acceleration approach", where one only tries to introduce some standard multigrid components into classical methods or into codes that are already available. For some examples, in particular the anisotropic convection-diffusion model operator and the (incompressible) Navier-Stokes equations, the gain that can be achieved by the acceleration approach is discussed.

With respect to multigrid literature, we generally refer to the multigrid bibliography [2]. We assume that the reader is familiar with the basic multigrid ideas as presented in [19], [3] or [8].

1. Introduction

A. Brandt generally claims that any problem that is suitable for an efficient multigrid (MG) treatment can be solved in less than 10 "work units" provided that the MG components are chosen correctly. On the other hand, users are often glad even if they succeed in reducing the computing time of their codes from 10,000 to 5,000 work units by incorporating some MG features.

How can that be? Who is right? Is A. Brandt wrong, far too optimistic with his postulate? Or are the users wrong, not being aware of what should be done, being satisfied with the choice of obviously bad MG components?

There has been a discussion going on about these questions for many years. Unfortunately, the discussion was often dominated by misunderstandings and emotions. In this short paper, we only want to illustrate some aspects of this discussion. For simplicity, we denote A. Brandt's view as the *"optimal MG approach"* and the users' view sketched above as the *"MG acceleration approach"*. In the optimal MG approach, the intension is to find a suitable discre-

---
[1] Suprenum GmbH, Bonn
[2] Gesellschaft für Mathematik und Datenverarbeitung mbH, St. Augustin
[3] University of Cologne

tization, optimal grid structures, and optimally taylored MG components for any given problem.

In the MG acceleration approach the typical situation is the other way round: A numerical code is already there, i.e. the discretization and the grid structures have been chosen before and a (classical) solution method is implemented. On the basis of these decisions, one tries to incorporate MG components into the code in order to accelerate it: MG components that fit to the chosen approach and that do not require too many changes in the program.

We will discuss the latter approach for some basic models in computational fluid dynamics (CFD) in the following sections. We want to point out, however, that we will only deal with "real" MG. So let us first make some introductory remarks about real MG and its development in the past.

Clearly, there have been and are many numerical approaches for PDEs or other mathematical models employing multiple grids (coarser and finer ones). For instance, coarse grids may be used in order to provide first approximations for fine grids, extrapolations are calculated by the use of coarser and finer grids, data compression and filtering are processes where fine grids may be replaced by coarse ones, etc.

Although all these and similar approaches may have certain features in common with the MG idea, they should, however, not be mixed up with real MG. Many such misunderstandings can indeed be found in the literature. Some of those misunderstandings are very obvious and therefore not dangerous. However, there are several more subtle ones which have had a significantly negative effect on the acceptance of multigrid methods.

Real MG consists of the combination of *smoothing* and *coarse grid approximation*. The real MG approach provides *methods* of solving elliptic PDEs and many other problems *where the number of (sequential) operations that have to be performed is proportional to the number of discrete unknowns (with a moderate factor of proportionality)*.

The MG idea and first theoretical results were published by Fedorenko and Bakvalov about 25 years ago. Although the essential theoretical properties of MG had already been recognized then, nobody was aware of the practical significance of this idea and the generality of the approach. So the possibilities contained in the MG idea remained undiscovered and contributions to this subject were neglected. It was 10 years later that A. Brandt became aware of these ideas and recognized their potential. He recognized the practical efficiency and the generality of MG methods and was the first one to apply the MG principle to CFD problems. However, as A. Brandt was not willing to go the normal way of strict mathematical proof, he had little success in drawing the numerical analysts' attention to the MG ideas in the beginning. And as he did not have the personnel resources to provide the CFD scene with real MG software products, he was not able to convince the CFD experts of the MG potential. After all, for the last 10 years an increasing community of numerical analysts, CFD experts, physicists, and other scientists have been engaged in the MG area. A breakthrough has been achieved, however, only during the last five years. Today, the idea has gained acceptance on a broad user front, although there are still arguments between mathematicians representing different MG directions and some uncertainty exists with respect to the limitations of the approach.

To come back to our initial topic - the main discussion goes on, however, about the *right* approach. We will consider the optimal MG versus the MG acceleration approach here for two typical situations.

The first situation is the one in which one tries to extend a classical single grid (SG) method to an MG method by providing an MG structure and using the SG method for smoothing. We consider this situation for the anisotropic convection-diffusion / Poisson operator in Section 2. We can recognize from our consider-ations here that the acceleration approach may indeed give a remarkable speed up in many cases. It has, however, to be applied correctly.

The second situation is characterized by a more complex CFD model. Here the alternatives are either to replace only a part of the algorithm by a standard MG solver or to use MG for the whole model. We discuss this for the semi-implicit versus the fully implicit treatment of the Navier-Stokes equations in Section 3. The speed up that can be achieved by the acceleration approach here typically is very limited.

2. From a single-grid to a multigrid method

A typical situation, particularly in CFD, is that programs are available and in use for certain applications. These codes usually are based on classical numerical (single-grid) methods often with sophisticated, optimized components (e.g. using optimal SOR- or ADI- parameters etc.). The user or the programmer of these codes then becomes aware of MG and gets an idea of how much faster he could solve his problem by the use of a suitable MG approach. He would like to exploit this possibility but is not willing to change the whole concept of his code and to entirely rewrite it. (This often would be not at all possible because of the com-plexity of the code.) Consequently, he wants to change only some parts of his code by incorporating MG components.

For iterative codes, one way of doing this, is to implement only the standard coarsening grid structure, to define the multigrid transfer operations, and to use V- (or F- or W-) cycles instead of the single-grid (SG) iteration. Here the SG iter-ation, which was supposed to converge towards the solution in the single-grid approach, plays the role of the smoothing procedure now. In many cases, this approach may work as many classical SG methods have - not optimal, but - rea-sonable smoothing properties. This approach is particularly promising if the pro-grammer is ready to change parameters, using *good smoothing parameters* instead of *optimal convergence parameters*. However, there are very important other cases where this procedure can basically provide no essential gain.

To illustrate this, let us consider the (linear) convection-diffusion operator with constant coefficients $c,$, d

$$A = \sum_{i=1}^{n} c_i \frac{\partial}{\partial x_i} - d \frac{\partial^2}{\partial x_i^2} \quad (n = 2 \text{ or } n = 3) \tag{1}$$

which may be regarded as a model operator for many fluid flow problems (e.g. for the development of smoothing techniques for the Navier-Stokes equations, cf. Section 3).

If d is not too small, the dominating part of the operator is of ∇^2 -type. In practice, however, often nonuniform grids have to be used which may lead to anisotropic operators:

If the meshsizes h_i are chosen differently in each direction the resulting difference operator is

$$A_h = \sum_{i=1}^{n} c_i \partial_{h_i} - d \partial_{h_i}^2$$

Here, ∂_{h_i} denotes a difference approximation for the first order derivatives, $\partial_{h_i}^2$ for the second order derivatives. Clearly, A_h is equivalent to and can be obtaine by a discretization of the differential operator

$$\tilde{A} = \sum_{i=1}^{n} c_i \gamma_i \frac{\partial}{\partial x_i} - d \gamma_i^2 \frac{\partial^2}{\partial x_i^2}$$

on a uniform grid with meshsizes $h_i \equiv h_1 = h$. The anisotropy is then generated by the meshsize ratios $\gamma_i = h_i/h_1$. The meshsizes may vary locally (see Figure 1) which leads to non-constant coefficients γ_i.

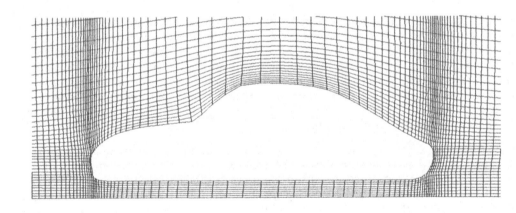

Figure 1. A typical grid for 2D-simulations of flow past cars. The local meshsize ratios h_1/h_2 of the cells vary enormously.

In order to discuss some cases in more detail, we first consider the main part of the above operator $\tilde{A}$ on a uniform 3D grid (meshsize h):

$$\tilde{A}_h = \sum_{i=1}^{3} a_i \partial_h^2 ; \quad (a_i = c_i \gamma_i)$$

We distinguish three cases

(1) $a_1 = a_2 = a_3 = 1$

(2) $a_1 = 100, a_2 = a_3 = 1$

(3) $a_1 = a_2 = 100, a_3 = 1$.

Case (1): If pointwise relaxation is used here as an SG method in order to solve the equation, the optimal SOR relaxation parameter results (for a 64^3 grid) as

$$\omega = \omega_{opt} \approx 1.91$$

(The corresponding convergence factor ρ and the factor $\rho^{1/WU}$ of error reduction per work unit[4]) are given in Table 1.)

If this SOR method (with $\omega = \omega_{opt}$) is used as a smoother in an MG context, this gives very unsatisfactory MG convergence (Table 1, $\omega = \omega_{opt}$ -column). It is well-known, however, that the standard RB-Gauss-Seidel method with $\omega = 1$ has nice smoothing properties and therefore gives good MG convergence (convergence factors $\rho \approx 0.2$, see Table 1 $\omega = 1$ -column). Used as an SG solution method on its own, Gauss-Seidel with $\omega = 1$ is, of course, absolutely out of the question.

	$\omega = 1$		$\omega = \omega_{opt} = 1.191$	
	ρ	$\rho^{1/WU}$	ρ	$\rho^{1/WU}$
SG(SOR)	0.997	0.997	0.906	0.906
MG	**0.229**	0.656	0.825	0.946

Table 1.   Asymptotic error reduction factors ρ and factors $\rho^{1/WU}$ of error reduction per work unit for pointwise Gauss-Seidel relaxation and the corresponding MG methods. The MG-cycles used are V-cycles with 1 relaxation before and 1 relaxation after the coarse grid correction. The relaxations are performed in a red-black ordering. All results refer to 64^3 grid points.

Case(2): In this case point relaxation makes no sense, neither for convergence nor for smoothing purposes (if standard coarsening is applied [20]). If one uses a relaxation method as an SG solver in this case, one usually would take linewise Gauss-Seidel relaxation (or LSOR). Such a linewise Gauss-Seidel relaxation (LSOR with $\omega = 1$) also turns out to be a good smoother, giving good MG convergence ($\rho \approx 0.1$).

A work unit is the computational work equivalent to one smoothing step on the finest grid. This naive definition is used in this paper for simplicity, although this definition may be somewhat problematic with respect to certain comparisons.

	point (RB, $\omega = 1$)		line (zebra, $\omega = 1$)	
	ρ	$\rho^{1/wu}$	ρ	$\rho^{1/wu}$
SG	0.997	0.997	0.887	0.887
MG	0.948	0.985	**0.081**	0.488

Table 2. Asymptotic error reduction factors ρ and factors $\rho^{1/wu}$ of error reduction per work unit for pointwise and linewise Gauss-Seidel relaxation and the corresponding MG methods. The MG-cycles used are V-cycles with 1 relaxation before and 1 relaxation after the coarse grid correction. The relaxations are performed in a RB and in a zebra-ordering, resp. All results refer to 64^3 grid points.

Case(3): In this case, none of the classical relaxation methods - neither SOR nor LSOR with *any* parameter - has satisfactory smoothing properties in connection with standard coarsening. MG solvers based on such bad smoothers are as inefficient as the relaxation methods themselves if these are used as SG solvers. (Even alternating line relaxation does not help.)

As has been shown in [20], *planewise* relaxation has to be used here for smoothing. The corresponding MG solver gives the typical good convergence ($\rho \approx 0.1$).

	point (RB, $\omega = 1$)		line (zebra, $\omega = 1$)		alternating line (zebra, $\omega = 1$)		plane (point 2D-MG, $\omega = 1$)	
	ρ	$\rho^{1/wu}$	ρ	$\rho^{1/wu}$	ρ	$\rho^{1/wu}$	ρ	$\rho^{1/wu}$
MG	0.966	0.990	0.942	0.983	0.914	0.987	**0.070**	0.701

Table 3. Asymptotic error reduction factors ρ and factors $\rho^{1/wu}$ of error reduction per work unit for MG using different relaxations for smoothing. The MG-cycles used are V-cycles with 1 relaxation before and 1 relaxation after the coarse grid correction. The linewise relaxations are performed in a zebra-ordering. Regarding the plane relaxation, see [20] for details. All results refer to 64^3 grid points.

We finally come back to the original operator (1). If the diffusion terms are small compared to the convection terms we have nearly purely convective flow where standard pointwise relaxation schemes again are not useful for smoothing. Possibilities in this case are

1. introduce artificial viscosity (isotropic or anisotropic) and use standard schemes like RB [3];

2. relax in flow direction (which is technically difficult) or use a global scheme like Symmetric Gauss-Seidel [3], [21];

3. if Symmetric Gauss-Seidel failes (as may occur in the case of Euler equations [13] use more powerful schemes like line-, plane- or ILU-relaxations.

3. Semi- or fully implicit multigrid treatment of the Navier-Stokes equations

As already sketched in the introduction, a different type of MG acceleration versus optimal MG can be illustrated in the case of the (incompressible) Navier-Stokes equations. We write the incompressible Navier-Stokes system of partial differential equations in operator formulation:

$$N\mathbf{u} + \nabla p = \mathbf{s} \tag{2}$$

$$\nabla \cdot \mathbf{u} = 0 \tag{3}$$

considering a domain $\Omega \subset \mathbb{R}^n$, ($n = 2$ or $n = 3$) with appropriate boundary conditions. $\mathbf{u}$ is the velocity field, p the pressure and $\mathbf{s}$ the external force. The (nonlinear) convection-diffusion operator

$$N = N(\mathbf{u}) = \rho\left(\sum_{i-1}^{n} u_i \frac{\partial}{\partial x_i}\right) - \mu\nabla^2$$

is applied to each velocity component u_i, where ρ and μ denote the density and the viscosity, respectively.

3.1 The semi-implicit approach

A well-known iterative solution method for the Navier-Stokes equations (in the above primitive variable $(\mathbf{u}, p)$ formulation) is the SIMPLE method (Semi-Implicit-Method for Pressure-Linked Equations [14]).

For the discretization, a staggered grid is used here with the velocity components u_i being stored on centers of cell faces and the scalar variable p being stored in the cell centers. (In contrast to the continuous variables, we mark the discrete variables by the subscript $_h$.)

One step of this method, calculating $\mathbf{u}_h^{n+1}$, p_h^{n+1} from $\mathbf{u}_h^n$, p_h^n, can be described in the following way:

1. *Prediction step* Evaluate an approximative velocity field $\bar{\mathbf{u}}_h$ using the old pressure p_h^n:

$$N'_h(\mathbf{u}_h^n)\bar{\mathbf{u}}_h + \nabla_h p_h^n = \mathbf{s}_h \tag{4}$$

Here $N'_h(\mathbf{u}_h^n)$ denotes some simple discrete linearization of the convection-diffusion operator N. In the original SIMPLE algorithm, usually a crude approximation $\bar{\mathbf{u}}_h$ is computed by applying only a few line-relaxations or some iterations of the Strongly Implicit Procedure (SIP) [18] to equation (4).

. As this approximation will not fulfil the discrete continuity equation, corrections $\delta\mathbf{u}_h$, δp_h are computed by use of the - once more simplified - equations

$$N_h^{\cdot}\delta\mathbf{u}_h + \nabla_h\delta p_h = 0 \tag{5}$$

$$\nabla_h \cdot \delta\mathbf{u}_h = -\nabla_h \cdot \mathbf{u}_h \tag{6}$$

Here $N_h^{\cdot}$ is a diagonal approximation of $N'_h(\mathbf{u}_h^n)$ and (6) means that

$$\bar{\mathbf{u}}_h + \delta\mathbf{u}_h$$

fulfils the discrete continuity equation. For the actual computation of δp_h , (5) is resolved:

$$\nabla_h \cdot \left((\dot{N}_h)^{-1}\nabla_h\delta p_h\right) = \nabla_h \cdot \bar{\mathbf{u}}_h$$

This so-called *pressure correction equation* (PC) obviously is a Poisson-like equation.

3. Finally $\delta\mathbf{u}_h$ is computed from (5):

$$\delta_h = -(\dot{N}_h)^{-1}\nabla_h\delta p_h$$

In order to ensure convergence, it is necessary to use underrelaxation in computing the new iterates

$$\mathbf{u}_h^{n+1} = \alpha_u(\bar{\mathbf{u}}_h + \delta\mathbf{u}_h) + (1 - \alpha_u)\mathbf{u}_h^n \ , \quad p_h^{n+1} = \alpha_p(p_h^n + \delta p_h) + (1 - \alpha_p)p_h^n$$

with relaxation parameters $\alpha_u, \alpha_p < 1$. Their optimal value depends on the Reynolds number.

The SIMPLE procedure described above is a very popular method, as it requires - apart from the prediction step - mainly the solution of the Poisson-like PC-equation in each iteration step.

The convergence of the SIMPLE method depends on several physical parameters which determine the coupling of the system (2) , (3) and on numerical features (meshsize h, approximation of N'_h by N_h, solution accuracy for the PC-equation etc.). The convergence may indeed become very slow.

Mathematically closely related to the SIMPLE approach is the semi-implicit approach for the time-dependent Navier-Stokes equations

(with $\rho\dfrac{\partial}{\partial t}\mathbf{u} + N\mathbf{u} + \nabla p = 0$ instead of (2)):

Here the semi-implicit approach within a time marching procedure allows time-step sizes Δt which are limited only by the velocity Courant number instead of the (usually much more restrictive) acoustic Courant number as in purely explicit schemes. An example of the semi-implicit method is the SOLA-code [9]. In the original SOLA code a pointwise SOR relaxation is used for the iterative solution of the PC-equation [4].

A typical MG acceleration approach for both, the steady state SIMPLE and the time-dependent SOLA, is to replace the (L)SOR or SIP solver for the PC-equation by a suitable MG solver. This is very attractive from the technical point of view: The structure of the CFD code is not changed, only one module has to be replaced by another. This involves changes in only a small piece of the usually large CFD code. One has to be aware, however, that this replacement will *not at all* change the limited overall convergence of the semi-implicit approach as far as this limitation is caused by the coupling of the Navier-Stokes system.

This MG acceleration approach has been studied by several authors. Results have been reported by [4], [6], [12] and others. We pick out just some of these results here that look typical to us. In [12] several numerical investigations and comparisons have been made. A 2D incompressible Navier-Stokes equation (with Reynolds number Re=2500) is solved using the SIMPLE approach. Within this approach, the PC-equation is solved by

- an alternating line relaxation as SG method

- an MG method

The work units needed to solve the considered problem on the basis of the SIM-PLE approach with SG and with MG, resp., are

grid	SG	MG	speed up
17^2	1200	1010	1.19
33^2	2940	2160	1.36
65^2	7380	4350	1.70

Table 4.

The overall speed up that is obtained here by replacing SG by MG for the *PC part of the code* is obviously ≤ 2.

In [4] a compressible 2D-Navier-Stokes problem is solved on the basis of SOLA. Again an SG and an MG approach are compared. Whereas for a 12^2-grid the MG acceleration gives a speed up of approximately 1.6, for a 24^2-grid the speed up is 3.4.

In [6], for a time-dependent incompressible 3D-problem with $28 \times 12 \times 140$ grid points (SOLA) a speed up of about 3.3 is gained by the use of MG instead of SG.

In general, the mere MG-solution of the PC equation is not the optimal way to exploit the full MG power in the numerical solution of the Navier-Stokes equations. Their coupling and the resulting possibly very slow convergence of the global iteration (or the time-step size restrictions in case of SOLA-type algorithms) will finally determine the overall computing time needed independently of how efficiently the PC-equation is solved. In particular, the acceleration approach can not lead to methods where the computational work is proportional to the number of unknowns on the finest grid.

3.2 The fully implicit approach

Theoretically more attractive - but also much more difficult to realize - is the fully implicit treatment of the coupled nonlinear system (2) , (3) by an MG method. The crucial point which has to be solved here is the development of reliable smoothing procedure. Experiences with the fully implicit treatment have been reported to our knowledge only for geometrically simple domains up to now and the main emphasis was put into the development of good smoothing procedures. Since an analytical two-level analysis for the nonlinear system is not available a lot of experimental work was necessary.

Different smoothing procedures have been investigated:

- Brandt proposes a distributive relaxation technique, as principally described in [3]. Its application to system (2), (3) has been discussed first in [5]. Let u_h^n, p_h^n be the last iterates. Each momentum equation is smoothed separately by relaxing $\tilde{N}_h$, a suitable approximate linearization of N_h, producing a new approximation $\tilde{u}_h$. Here $\tilde{N}_h$ is applied to each single equation and is identical to the convection-diffusion operator which has been considered in Section 2. Therefore, the smoothing of $\tilde{N}_h$ should be oriented on the smoothing strategies for the scalar convection-diffusion operator as discussed in Section 2. The subsequently described distributive relaxation step then couples the velocity and pressure changes in such a way that the continuity equation is satisfied and the momentum equation (2) defect remains unchanged: The corrections δu_h and δp_h are obtained from

$$\delta \mathbf{u}_h = \nabla_h \chi_h, \quad \delta p_h = - \tilde{N}_h \chi_h$$

where χ_h is a "ghost" variable (c.f. [3]) defined at all cell centers being determined by relaxing the equation

$$\nabla_h^2 \chi_h = - \nabla_h \cdot \tilde{\mathbf{u}}_h$$

The new iterates

$$\mathbf{u}_h^{n+1} = \tilde{\mathbf{u}}_h + \delta \mathbf{u}_h, \quad p_h^{n+1} = p_h^n + \delta p_h$$

leave the momentum defect unchanged as long as the operators $\tilde{N}_h$ and ∇_h commute. This is, however, not true in general. An improved distributive scheme which does not assume commutativity has been proposed by Fuchs [7].

- Similar to the approach discussed in Section 2, iterative single-grid solution schemes can be used as smoothers. An obvious choice is the SIMPLE-procedure as described in 3.1. Lonsdale [12] proposes a smoothing scheme based on the extended pressure correction scheme of VanDoormaal and Raithby (a SIMPLE variant). One smoothing step is one SIMPLE-iteration with a particular choice of N_h^*. This scheme is also a distributive relaxation but does not assume commutativity of the operators. However, compared to Brandt's relaxation it is more expensive in terms of computational work.

  Following the discussion in Section 2 concerning the role of relaxation parameters, the effect of the underrelaxation parameters used in the SIMPLE method has to be investigated. Whereas the convergence of the SIMPLE as SG method depends heavily on these parameters the smoothing properties seem to be more unsensitive to the parameter choice [1].

- Distributive relaxations work well as long as the coefficients in the linearized operator N'_h vary smoothly. If the coefficients vary rapidly (on the scale of the mesh size) as it may occur in case of internal flows or on coarser grids, a *box* relaxation scheme may be preferable. Such a scheme has been implemented by Vanka [21] on the basis of a symmetric coupled Gauss-Seidel relaxation. (Here all $(2n + 1)$ cell variables are updated simultaneoulsy by inverting a $(2n + 1)^2$ matrix in each cell.) Clearly, this scheme requires more computational work per smoothing step than the distributive schemes and it depends on the particular applications whether the increased smoothing effort pays in terms of faster MG-convergence.

With this box relaxation, Vanka indeed obtains for incompressible 2D Navier-Stokes equations an MG convergence that is essentially independent of the meshsize. This means that the number of operations is really proportional to the number of grid points. The computational work increases, however, substantially if the Reynolds number becomes larger. Vanka needs the following work units for a 321^2-grid

Re	100	400	1000	2000
WU	34	32	78	131

Table 5.

An optimal MG approach in the sense of Brandt is a combination of the fully implicit smoothers described above and the Full multigrid (FMG) algorithm. FMG [3], [19] is a non-iterative MG method which - under certain assumptions - ensures that the error of the MG solution is of the same size as the discretization error. Using optimal components and regarding the FMG accuracy as sufficient, Brandt demonstrates in [5] that the 2D incompressible steady-state Navier-Stokes equations (for small Reynolds numbers and simple geometry) are really solvable in less than 10 WUs.

Re	8	100	500
WU	7.2	7.9	9.7

Table 6.

3.3 Other CFD models, some final remarks

Clearly, apart from the formulation in primitive variables (u, p) there are other possibilities to treat the incompressible Navier-Stokes equations. In 2D and certain symmetric 3D cases, a *stream-function-vorticity* formulation is useful. In this formulation, the pressure is eliminated and can be calculated separately by solving a Poisson equation. In [11], an efficient MG solver for the pressure equation in spherical coordinates was presented. We pointed out there, that the use of pure Dirichlet boundary conditions for the Poisson equation - as proposed and tried by several authors - *must* lead to convergence troubles as the Dirichlet problem is ill-posed here. (This shows impressively that MG reacts very sensitively if the acceleration approach is not performed correctly.)

An MG treatment of the full Navier-Stokes model of an incompressible fluid in a rotating spherical gap was investigated by Linden in [10].

We want to make only few remarks about the MG treatment of the *compressible* Navier-Stokes equations. This treatment can either based on the incompressible methods or, in case of very high Reynolds numbers, on MG methods for the Euler equations as they have been developed by Jameson, Hemker and others. Regarding the latter approach, an overwiew is given in [16]. For medium Reynolds numbers Brandt [3] gives an extension of the distributive relaxation scheme to compressible equations which involves a simultaneous relaxation of the scalar variables pressure, density and temperature. On the basis of these ideas, an MG code for the time-dependent compressible Navier-Stokes problem was developed [15] in our MG group. It simulates the gas flow in a combustion chamber of a reciprocating engine. The Reynolds number was rather low (Re $\approx$ 70) due to turbulence effects. Convergence factors obtained for a 2D model with this relaxation are given in Table 7

grid	ρ	$\rho^{1/WU}$	WU
32^2	0.25	0.60	9.5
64^2	0.30	0.63	10.4

Table 7.   Convergence factors and work units for the (time-dependent) compressible Navier-Stokes equations. A distributive relaxation scheme (see [15]) has been used as a smoother within a fully implicit time discretization. The work units refer to one time-step.

In order to model the combustion chamber more realistically, a 3D model was developed using a Cartesian grid with a special boundary approximation [17]. Depending on the crank angle position a large anisotropy had to be taken into account. According to the discussion in Section 2, linewise relaxation was used and MG convergence factors of less than 0.25 were achieved.

We finally would like to point out that the models we have discussed in Section 2 and Section 3 are to be regarded only as representative for a much larger class of situations where our considerations apply. In particular, for real models the considerations conceiving anisotropy (Section 2) and those conceiving Navier-Stokes smoothing procedures (Section 3.2) have to be combined. We do not want to discuss these technically complicated situations here.

There is no concrete conclusion we may draw from our considerations in this short note and no general recommendation to use or not to use the acceleration approach. We have seen that the acceleration approach - *if applied correctly* - may give a remarkable speed up compared to classical methods. In other cases, the gain is small.

Clearly, there is a trade-off between the saving of computing time and the effort in reprogramming large CFD codes. This trade-off should be evaluated before the decision acceleration/optimal approach is made.

In any case, it should be regarded as a task also for those MG experts who usually prefer to work in the optimal MG approach not to fight against the acceleration approach, but to contribute to answer the question where and why the acceleration approach is useful and where it is not. Furthermore, the MG experts should help to develop efficient acceleration ideas and should conduct their realization. Finally, they should actively contribute to find practically relevant compromises with respect to the trade-off mentioned above.

4. References

[1]   Barcus, M.
        Berechnung zweidimensionaler Strömungsprobleme mit Mehrgitterverfahren.
        Diplomarbeit Universität Erlangen-Nürnberg, 1987.

[2]   Brand, K., Lemke, M., Linden, J.
        Multigrid bibliography.
        Arbeitspapier der GMD 206, Gesellschaft für Mathematik und Datenverarbeitung,
        Bonn 1986.

[3]   Brandt, A.
        Multigrid techniques: 1984 guide with applications to fluid dynamics.
        GMD-Studie 85, Gesellschaft für Mathematik und Datenverarbeitung, Bonn 1984.

[4]   Brandt, A., Dendy, J.E., Ruppel, H.
        The multigrid method for semi-implicit hydrodynamics codes. J. Comp. Phys. **34**,
        348-370, 1980.

[5]   Brandt, A., Dinar, N.
        Multigrid solutions to ellitpic flow problems.
        In: Numerical Methods for Partial Differential Equations (S.V. Parker ed.), 53-147,
        Academic Press, New York, 1979.

[6]   Brockmeier, U., Mitra, N.K., Fiebig, M.
        Implementation of multigrid in SOLA.
        Notes on Numerical Fluid Mechanics, **13**, 23-30, Vieweg 1986.

[7]   Fuchs, L.
        Multi-grid schemes for incompressible flows.
        Notes on Numerical Fluid Mechanics, **10**, 38-51, Vieweg 1984.

[8]   Hackbusch, W.
        Multi-grid methods and applications.
        Springer Series in Computational Mathematics, **4**, Springer 1985.

[9]   Hirt, C.W., Nichols, B.D., Romero, N.C.
        SOLA - a numerical solution algorithm for transient fluid flows.
        LASL report LA-5652, 1975.

[10]   Linden, J.
        Mehrgitterverfahren für das erste Randwertproblem der biharmonischen Glei-
        chung und Anwendung auf ein inkompressibles Strömungsproblem.
        GMD-Bericht 164, Oldenbourg 1986.

[11]   Linden, J., Trottenberg, U., Witsch, K.
        Multigrid computation of the pressure of an incompressible fluid in a rotating
        spherical gap.
        Notes on Numerical Fluid Mechanics, **5**, 183-193, Vieweg 1982.

[12]   Lonsdale, G.
        Solution of a rotating Navier-Stokes problem by a nonlinear multigrid algorithm.
        Numerical Analysis Report 105, University of Manchester, 1985.

[13]   Mulder, W.A.
        Analysis of a multigrid method for the Euler equations of gas dynamics in two
        dimensions.
        Proceedings of the 3rd International Copper Mountain Conference on Multigrid
        Methods, 1987.

[14]   Patankar, S.V., Spalding, D.B.
       A calculation procedure for heat, mass and momentum transfer in three-dimen-
       sional parabolic flows.
       Int. J. Heat Mass Transfer, **15**, 1787-1806, 1972.

[15]   Ruttmann, B., Solchenbach, K.
       A multigrid solver for the computation of in-cylinder turbulent flows in engines.
       Notes on Numerical Fluid Mechanics, **10**, 87-108, Vieweg 1984.

[16]   Schröder, W., Hänel, D.
       A comparison of several MG-methods for the solution of the time- dependent
       Navier-Stokes equations. In: Multigrid Methods II (W. Hackbusch, U. Trottenberg
       eds.),
       Lecture Notes in Mathematics 1228, Springer 1986.

[17]   Solchenbach, K., Steckel, B.
       Numerical simulation of the flow in 3D-cylindrical combustion chambers using
       multigrid methods.
       Arbeitspapiere der GMD 216, Gesellschaft für Mathematik und Datenverarbeitung,
       Bonn 1986.

[18]   Stone, H.L.
       Iterative solution of implicit approximations of multidimensional partial differential
       equations.
       SIAM J. Num. Anal., **5**, 530-560, 1968.

[19]   Stüben, K., Trottenberg, U.
       Multigrid methods: Fundamental algorithms, model problem analysis and applica-
       tions. In: Multigrid Methods (W. Hackbusch, U. Trottenberg eds.),
       Lecture Notes in Mathematics 960, Springer 1982.

[20]   Thole, C.A., Trottenberg, U.
       Basic smoothing procedures for the multigrid treatment of elliptic 3D-operators.
       Notes on Numerical Fluid Mechanics, **11**, 102-111, Vieweg 1984.

[21]   Vanka, S.P.
       Block-implicit multigrid solution of Navier-Stokes equations in primitive variables.
       J. Comp. Phys., **65**, 138-158, 1986.

An Algebraic Approach to Performance Analysis

Dieter Müller-Wichards

German Aerospace Research Establishment (DFVLR)
Bunsenstrasse 10
D-3400 Göttingen

1. Introduction

With the availability of supercomputers for large and computationally intensive applications during the last decade a new era of performance evaluation activities was started. These activities were especially motivated by the wide range of performance data, achievable on a single machine, bounded from above only by the advertized peak rates.

A central issue in performance investigations on supercomputers involves the question whether there is an inherent pairing between architectures and applications. Choices are made at each stage of the development of an application: starting at the top level with the mathematical model, proceeding with the selection of the algorithm and the numerical method, and ending with a particular implementation at the bottom level. All these decisions may have a considerable effect on the eventual performance. This has to be taken into account if we want the statement "application A has performance r on machine M" to be meaningful. The behaviour of an application on a specific machine is not properly understood before prediction (through a performance model) and measurement have been brought to to an agreement, at least to a certain extent.

Current practice and possible directions are summarized in [15].

In this paper we present an approach to performance modelling that is both, analytic and synthetic: analytic in the sense that the parallel-sequential structure of an application can be specified in a manner of stepwise refinement, synthetic in the sense that the total performance can be calculated, once the performance of each building block is specified. The value, thus obtained, often reflects parameters of the machine organization and of the representation of the application. These parameters can be used for a sensitivity analysis that may reveal the contribution of various factors to the performance behaviour of an application.

The basic algebraic setting we use is that of a "performance algebra" $\wp$, of an "application algebra" $\mathscr{H}$, and of a mapping $\varphi : \mathscr{H} \rightarrow \wp$ that is compatible with the algebraic operations in $\mathscr{H}$ and $\wp$. $\mathscr{H}$ and $\wp$ turn out to be ordered semigroups, where arithmetic operations, data movements, and delays (such as dispatching overhead) are algebraically modelled.

By exploiting the properties of our algebraic framework we can retrieve a number of well known laws, among them Amdahl's law, the law of the harmonic mean, and

---

Part of this work was performed during a sabbatical at the IBM T.J.Watson Research Center, Yorktown Heights, New York, USA.

Hockney's $n_{1/2}$ and $s_{1/2}$ laws, but also a number of additional results, such as an "outer loop parallelization inequality".

We examine the algebraic properties of an "implementation mapping" and formulate an "abstract benchmarking problem" in algebraic terms. Proofs can be found in [16].

In the section on examples we study a simple array operation, model the effect of memory hierarchies on the performance of a matrix multiplication, and discuss a macro-pipelined Gauß-Seidel Iteration for a local-memory multiprocessor and for a vector machine.

2. The Performance Algebra

In this section we shall define the performance algebra and summarize a number of its algebraic and analytical properties. We have chosen performance and operation count as the first two components of the elements. This choice is somewhat arbitrary and primarily a matter of taste; in fact, the selection of time and operation count would have led to an isomorphic algebraic structure. Nevertheless we feel that performance as a primary entity can easily be related to the nominal performance of a processor and thus give an indication how well the application was mapped to the processor.

Let $\mathbb{R}_+$ denote the set of positive real numbers and let

$$\wp' := \mathbb{R}_+ \cup \{\infty\} \times \mathbb{R}_+ \times \{0,1\}$$

$$V := \mathbb{R}_+ \cup \{\infty\} \times \{i\} \times \{0\} \text{ where } i := \sqrt{-1}$$

$$D := \mathbb{R}_+ \cup \{\infty\} \times \mathbb{R}_+ \times \{0\}$$

$$C := \mathbb{R}_+ \cup \{\infty\} \times \mathbb{R}_+ \times \{1\}$$

The set we shall primarily deal with in this paper is $\wp := C \cup D \cup V$. The elements of $\wp$ are triples where the first entity represents a performance, the second an amount of work, and the third contains information to determine what type of element it is, i.e. whether it belongs to V, D, or C. To the elements of the subset V we shall refer to as "delay elements" (notice that the amount of work assigned to delay elements is imaginary), similarily to D as "data movement elements", and to C as "arithmetic elements". Obviously $\wp' = C \cup D$ the elements of which we shall refer to as "proper elements".

Having described the objects we want to deal with we shall now concentrate on defining the operation or rather a parametric family of operations with respect to these objects.

Definition:

Let $1 \leq q \leq \infty$ then we construct the operation $\otimes^q$ as follows:

First of all we define a mapping $\sigma: \wp \rightarrow \mathbb{R}_+ \cup \{0\}$ as $\sigma(g) := \dfrac{|w|}{r}$ for all

$g = (r,w,u) \in \wp$. Let $g_1 \in \wp'$ and $g_2 \in \wp$ then the element $g_1 \otimes^q g_2$ has the following components:

$$u := u_1 \vee u_2$$

$$w := Re\,(u_1 w_1 + u_2 w_2 + \neg u(w_1 + w_2))$$

$$r := \frac{|w|}{\left(\sigma(g_1)^q + \sigma(g_2)^q\right)^{\frac{1}{q}}}\,, \quad q < \infty$$

$$r := \frac{|w|}{\max(\sigma(g_1),\,\sigma(g_2))}\,, \quad q = \infty$$

As obviously $\otimes^q$ is commutative where we have defined it, it suffices to explain the operation on V. Let $g_1, g_2 \in V$ then the resulting amount of work we define as $w := i$ whereas the calculation of u and r is done in the same manner as in the previous case.

Obviously σ can be interpreted as a function that assigns to each element of $\wp$ the corresponding "time".

The resulting amount of work for $g = g_1 \otimes^q g_2$ is summarized in the following table, where operands are taken from various combinations of subsets of $\wp$.

$g_2 \backslash g_1$	V	D	C
V	i	w_1	w_1
D	w_2	$w_1 + w_2$	w_1
C	w_2	w_2	$w_1 + w_2$

The definition for the resulting amount of work reflects the fact that in practice no particular award is granted for moving large numbers of data around when e.g. MEGAFLOPS are calculated.

In the sequel we shall deal with *ordered semigroups*, i.e. associative algebraic structures that are equipped with a partial order, compatible with the semigroup operation. Semigroups that contain an identity are called *monoids*. Transitive and reflexive relations that do not have the antisymmetry property are called *preorders*. We speak of *full* orders if each pair of elements can be compared. Finally a semigroup is called *strictly* ordered if a strict inequality remains strict when both sides are multiplied by an element of the semigroup. Details can e.g. be found in [5] or [13].

With these definitions in mind we can now introduce a relation $\leq$ on $\wp$ in the following way.

Definition:

g_1, g_2 are called *comparable*, if either $g_1, g_2 \in V$ or $g_1, g_2 \in D$ or $g_1, g_2 \in C$.

Let $g_1, g_2 \in \wp$ then $g_1 \leq g_2$, if g_1 and g_2 are comparable and either $\sigma(g_2) < \sigma(g_1)$ or $\sigma(g_2) = \sigma(g_1)$ and $r_2 \geq r_1$.

Thus the relation $\leq$ can be interpreted as a lexicographical order on the set of pairs $(-\sigma(g), r)$.

Definition:

Let $g_1, g_2 \in \wp$ then $g_1 \underset{\sim}{\leq} g_2$, if g_1 and g_2 are comparable and $\sigma(g_2) \leq \sigma(g_1)$.

We restrict comparability to element pairs of the same class (i.e. V, D, or C) because intuitively it does not make sense to compare e.g. data movements with arithmetic operations. Moreover, compatibility of the order relations defined above with the semigroup operations would be lost if we were to attempt comparisons across class boundaries.

Both order relations are based on "time" as the primary (or only) criterion, rather than on performance (alone). The reason is that quite frequently, when algorithmic choices are made for applications on vector or parallel computers, higher operation counts or even slower convergence rates are traded in for higher performance. Time seems to be a fair measure as to which decision should be made.

The relationship between the two orderings can be summarized in the following way:

Let $g_1, g_2 \in \wp$ then the following assertions hold :

1. If $g_1 \leq g_2$ then $g_1 \underset{\sim}{\leq} g_2$.

2. If $g_1 < g_2$ with respect to $\underset{\sim}{\leq}$ then $g_1 < g_2$ with respect to $\leq$.

3. If the amount of work for both elements is equal then both relations have identical meaning.

The following theorem describes the over all behaviour of the algebraic structure that is our object of investigation in this section.

Theorem 1

$1 \leq q < \infty$, $(\wp, \otimes^q)$ is a partially ordered monoid with respect to the relation $\leq$.

For $q = \infty$, $(\wp, \otimes^q)$ is a partially preordered monoid with respect to the relation $\underset{\sim}{\leq}$.

The identity element is $e = (\infty, i, 0)$.

■

The weaker assertion for $\otimes^\infty$ is due to the fact that the inequality $g_1 \leq g_2$ is not always preserved if we multiply both sides with respect to $\otimes^\infty$ as the following counterexample shows. In fact, multiplication with a "dominating" element g with the property $\sigma(g) \geq \max(\sigma(g_1), \sigma(g_2))$ can reverse the inequality for the products :

Let $\sigma(g_1) > \sigma(g_2)$ and $w_1 > w_2$, hence $g_2 > g_1$.

However

$$r(g_1 \otimes^\infty g) = \frac{w + w_1}{\sigma(g)} > \frac{w + w_2}{\sigma(g)} = r(g_2 \otimes^\infty g)$$

It is important to note that our failure to establish compatibility of the relation $\leq$ with the operation $\otimes^\infty$ cannot simply be repaired by chosing a "better" second criterion for the comparison. In fact, it can be shown that any criterion second to "time" (if we stick to "lexicographical type" orderings) used to guarantee anti-symmetry leads to the same failure ([3]).

Subsequently we shall implicitly exploit associativity when dealing with finite pro-ducts without worrying about parentheses. In addition to the global algebraic structure of $\wp$ we shall now describe its substructure.

Definition:

A nonempty subset B of a semigroup A is called a **subsemigroup** of A if $B \bullet B \subset B$. A nonempty subset C of A is called an **ideal** if $C \bullet A \subset C$.

Theorem 2

1. For $q < \infty$ the sets C,D and V are strict fully ordered subsemigroups of $\wp$ with respect to $\leq$. For $q = \infty$ they are fully preordered with respect to $\underset{\sim}{\leq}$.

2. C is an ideal.

∎

For $g_i \in \wp$, $i = 1, \dots ,n$ we shall frequently write $\prod_{i=1}^{n}{}^{(q)}g_i$ to denote the n-fold prod-uct $g_1 \otimes^q \dots \otimes^q g_n$.

Furthermore, if all $g_i = g$ then we shall use the notation $n \otimes^q g$ for the n-fold prod-uct.

For those cases, where it is explicitly known that $q = 1$ we shall use $\oplus$ instead of $\otimes^1$ and $\sum^{(1)}$ instead of $\prod^{(1)}$. Similarily for $q = \infty$ we shall use $\otimes$ instead of $\otimes^\infty$.

Remark "Scalar Multiplication"
The "multiplication" of an element of $\wp$ with a natural number as introduced above can be generalized to arbitrary nonnegative numbers in the following way

$$\lambda \otimes^q g : = (\lambda^{1 - \frac{1}{q}} \cdot r, \lambda w, u) \text{ for } g \text{ proper}$$

and

$$\lambda \otimes^q g : = (\lambda^{-\frac{1}{q}} \cdot r, i, u) \text{ for } g \in V$$

provided $\lambda > 0$ whereas for $\lambda = 0$ we define $0 \otimes^q g := e$.

The scalar multiplication as we have defined it is, of course, something like the "exponentiation" of elements of $\wp$. It is merely for typographical reasons that we have preferred to write this operation as a "multiplication".

Let λ and μ be positive real numbers, then a couple of *associative* and *distributive* *laws* hold:

$$\lambda \otimes^q (\mu \otimes^q g) = (\lambda \cdot \mu) \otimes^q g$$

$$(\lambda + \mu) \otimes^q g = (\lambda \otimes^q g) \otimes^q (\mu \otimes^q g)$$

$$\lambda \otimes^p (g_1 \otimes^q g_2) = (\lambda \otimes^p g_1) \otimes^q (\lambda \otimes^p g_2)$$

Furthermore scalar multiplication has an *order preserving* property:

$$g \le h \;\Rightarrow\; \lambda \otimes^q g \le \lambda \otimes^q h$$

∎

Remark: Averaging and the Law of the Harmonic Mean

Let $H \in \{C,D,V\}$ and $g_j \in H, j = 1, \dots, n$ then we can define an "average" element $\bar{g}$ in the following way :

$$\bar{g} := (\frac{1}{n}) \otimes^q \prod_{i=1}^{n}{}^{(q)} g_i \quad \text{which has the property} \quad n \otimes^q \bar{g} = \prod_{i=1}^{n}{}^{(q)} g_i$$

For the performance of $\bar{g}$ we obtain:

$$\bar{r} = \frac{|\bar{w}|}{\sigma(\bar{g})} = \frac{1}{\left(\dfrac{1}{n} \displaystyle\sum_{i=1}^{n} \left(\dfrac{w_i}{\bar{w}} \dfrac{1}{r_i} \right)^q \right)^{\frac{1}{q}}}$$

which for $q = 1$ is nothing but the law of the harmonic mean as proposed in [18].
∎

Remark: Continuity of $\otimes^q$

Let the complex numbers be equipped with the usual and the set $\{0,1\}$ with the discrete topology and $\wp$ with the induced product topology.

The investigation of continuity of $\otimes^q$ can thus be reduced to regarding sequences within the subsemigroups V,C,D, i.e. let $H_i \in \{V,C,D\}, i = 1,2$ and $(g_i^n)_{n=1}^{\infty} \subset H_i$ with $\lim_{n \to \infty} g_i^n = g_i = (r_i, w_i, u_i)$. Let further $h^n := g_1^n \otimes^q g_2^n$ then it is a straightforward matter to verify that $\lim_{n \to \infty} h^n = g_1 \otimes^q g_2$.
∎

The following three theorems deal with the relationship between expressions that involve operations $\otimes^q$ for different q.

Theorem 3:

Let $g_i \in \wp, i = 1, \dots, n$ and $g^{(q)} := \prod_{i=1}^{n}{}^{(q)} g_i$.

Then, if $1 \le q \le p \le \infty$ we have the following inequalities:

$$g^{(q)} \le g^{(p)}$$

$$\lambda \otimes^p g^{(q)} \geq \lambda \otimes^q g^{(p)}$$

for an arbitrary number $\lambda \geq n$. Furthermore we have :

$$\lim_{q \to \infty} g^{(q)} = g^{(\infty)}$$

∎

In fact $g^{(q)}$ is a continuous and monotonous "trajectory" through $\wp$ on $[1, \infty]$, whose endpoints stand for sequential and parallel performance respectively.

Thus it may be legitimate to ask, how $g^{(q)}$ can be interpreted in terms of $g^{(1)}$ and $g^{(\infty)}$, in other words, what percentage sequentialism and parallelism is contained in $g^{(q)}$?

One can attempt an approximate answer to this question, by looking at the inequality

$$\left| \sigma(g^{(q)}) - \left\{ \lambda(q) \dot{\sigma}(g^{(1)}) + (1 - \lambda(q)) \sigma(g^{(\infty)}) \right\} \right| \leq \lambda(q) f(q) + (1 - \lambda(q)) h(q) =: u(q)$$

where $0 \leq \lambda(q) \leq 1, \lambda(1) = 1$ and $\lambda(\infty) = 0$.

Furthermore

$$f(q) := \left(1 - n^{\frac{1}{q} - 1} \right) \sigma(g^{(1)})$$

and

$$h(q) := \left(n^{\frac{1}{q}} - 1 \right) \sigma(g^{(\infty)}).$$

Obviously we have $u(1) = u(\infty) = 0$.

If we minimize u with respect to some functional on an appropriate function space for all functions λ with the described restrictions, we can at least give a best answer to the "percentage" question in a certain sense.

If we chose the functional

$$J(\lambda) := \int_1^\infty (u^2 + u'^2) \, dt$$

we end up with a special case of a Sturm-Liouville boundary value problem.

In the next theorem we derive an inequality for the "transposition" of expressions, where the "inner" and the "outer" operations are exchanged. This theorem serves as a basis for a corresponding theorem for parallel-sequential expressions in the next section.

Theorem 4: "Outer Loop Parallelization Inequality"

Let $1 \leq q \leq p \leq \infty$ and $g_{ij} \in \wp, i = 1, \dots ,m, \ j = 1, \dots ,n$ then

$$\prod_{j=1}^n {}^{(p)} \left(\prod_{i=1}^m {}^{(q)} g_{ij} \right) \geq \prod_{i=1}^m {}^{(q)} \left(\prod_{j=1}^n {}^{(p)} g_{ij} \right)$$

∎

It may be interesting to note that the proof makes use of a geometric inequality (Minkowski's Inequality for $l^{p/q}$ space).

As a byproduct we obtain an "associative inequality" if $\wp$ is equipped with $\otimes^q$ and $\otimes^p$,

Corollary 1: "Associative Inequality"

Let $1 \leq q \leq p \leq \infty$ and $h_i \in \wp$, $i = 1, \ldots ,3$ then

$$h_1 \otimes^p (h_2 \otimes^q h_3) \geq (h_1 \otimes^p h_2) \otimes^q h_3$$

∎

The following theorem provides a more or less coarse performance estimate for the execution of n parallel tasks on $p \leq n$ processors, circumventing the complexity of scheduling problems (see e.g. [17]).

Theorem 5:

Let π be a partition of the set $I := \{1, \ldots ,n\}$ into p classes, i.e. $\pi := \{I_1, \ldots , I_p\}$, let further $g_i \in \wp'$, $i = 1, \ldots ,n$,

$$g_\pi := \prod_{j=1}^{p}{}^{(\infty)} \left(\sum_{i \in I_j}{}^{(1)} g_i \right) \quad \text{and} \quad g^{(q)} := \prod_{i=1}^{n}{}^{(q)} g_i.$$

Then for every π there is a unique $q \in [1, \infty]$ such that $g_\pi = g^{(q)}$, where $q = 1$ if $p = 1$ and $q = \infty$ if $p = n$.

Let $|\pi| := \max(|I_j|, \ j = 1, \ldots ,p)$ and $q(\pi) := \log n / \log |\pi|$ then $q(\pi) = 1$ for $p = 1$ and $q(\pi) = \infty$ for $p = n$ and we have the following estimates:

$$\frac{1}{p} \, \sigma(g_{seq}) \leq \sigma(g_\pi) \leq |\pi| \, \sigma(g^{(\infty)})$$

$$\frac{1}{p} \, \sigma(g_{seq}) \leq \frac{|\pi|}{n} \sigma(g_{seq}) \leq \sigma(g^{q(\pi)}) \leq |\pi| \, \sigma(g^{(\infty)})$$

$$|\pi|^{-\frac{\log p}{\log n}} \sigma(g^{q(\pi)}) \leq \sigma(g_\pi) \leq |\pi|^{\frac{\log \frac{n}{|\pi|}}{\log n}} \sigma(g^{q(\pi)})$$

where $g_{seq} := g^{(1)}$.

The factor $|\pi|^{-\frac{\log p}{\log n}}$ in the last inequality is equal to 1 for $p = 1$ and $p = n$. The same is true for the other factor.

Remark:
If $\sigma(g_i) = \sigma$, $i = 1, \ldots , n$ then we have $\sigma(g_\pi) = \sigma(g^{q(\pi)}) = |\pi| \sigma$.

∎

In the following theorem we will restate Amdahl's law [1] in our algebraic setting and state a few additional results. There are various ways of interpreting g_1 and g_2 in the subsequent definition. Among the possible interpretations are the scalar and vector performance of a uniprocessor ($u = 1$) or the transfer rate of data from memory and from cache ($u = 0$), as described e.g. in [11].

Theorem 6 *"Amdahl's Law"*

Let $g_i = (r_i, 1, u)$, $i = 1,2$ and $g_1 < g_2$. and let λ be the fraction of work attributed to g_1 with $0 \le \lambda \le 1$. Let further $q \in [1, \infty]$ be selected such that $\otimes^q$ describes the appropriate type of parallelism between the work corresponding to g_1 and g_2. Then the effective performance can be stated as the "convex combination"

$$g^{(q)}(\lambda) := (\lambda \oplus g_1) \otimes^q ((1 - \lambda) \oplus g_2)$$

∎

It turns out that the problem of determining the optimal load distribution w.r.t. performance leads to the geometric problem of finding the minimum norm element of a hyperplane in $\mathbb{R}^2$.

Let $g_{max}^{(q)} := \max_{\lambda} g^{(q)}(\lambda)$ then the maximum is attained at

$$\lambda_{max}^{(q)} = \frac{r_1^p}{r_1^p + r_2^p} \quad \text{where } p = \frac{q}{q-1} \quad \text{for } 1 < q < \infty$$

Furthermore

$$r_{max}^{(q)} := r\left(g_{max}^{(q)}\right) = (r_1^p + r_2^p)^{\frac{1}{p}}$$

Both $r_{max}^{(q)}$ and $\lambda_{max}^{(q)}$ are continuous functions on $[1, \infty]$.

Remark:
Amdahl's law can be iterated to describe the performance in a somewhat more complicated environment, as e.g. in a vector machine with p processors:

Let g_0 and g describe the sequential and parallel performance for unit work, and λ the fraction of sequential work then

$$g_{eff} = (\lambda \oplus g_0) \oplus ((1 - \lambda) \oplus g)$$

Let further

$$g = \prod_{i-1}^{p}{}^{(\infty)} (\lambda_i \oplus g_i)$$

where g_i denotes the performance on the i-th processor and λ_i the fraction of parallel work assigned to it.

Finally let $g_i = (\mu_i \oplus g_i^s) \otimes^q ((1 - \mu_i) \oplus g_i^v)$, $i = 0,1, \ldots ,p$, where μ_i describes the scalar fraction of work and g_i^s the scalar and g_i^v the vector performance for unit work. Then, by virtue of the laws of scalar multiplication, the total performance can be written as:

$$g_{eff} = \left((v_0^S \oplus g_0^S) \otimes^q (v_0^V \oplus g_0^V)\right) \oplus \prod_{i-1}^{p}{}^{(\infty)}\left((v_i^S \oplus g_i^S) \otimes^q (v_i^V \oplus g_i^V)\right)$$

where

$$v_i^S = (1 - \lambda)\, \lambda_i\, \mu_i,$$

$$v_i^V = (1 - \lambda)\, \lambda_i\, (1 - \mu_i), \quad i = 1, \ldots, p,$$

and

$$v_0^S = \lambda \mu_0, \quad v_0^V = \lambda(1 - \mu_0),$$

and where we also have

$$\sum_{i=0}^{p} (v_i^S + v_i^V) = 1$$

∎

Remark:
The mathematical methods for determining the maximum performance and the corresponding work load distribution can be applied for any finite number of processors (or memory levels) with essentially the same results. If not all operations $\otimes^q$ have the same q the the maximum can be determined by working ones way up in the expression tree.

∎

3. The Application Algebra

We now turn our attention to an abstraction of applications and their representation as parallel / sequential structures. In mathematical terms the application algebra $\mathcal{H}$ is a *free semigroup* over a finite alphabet B with respect to the operations $\oplus$ and $\otimes$.

B can be chosen as an irreducible set (w.r.t. $\oplus$ and $\otimes$) of "basic algorithms". An expression of the type $A_1 \oplus A_2$ can be interpreted as sequential, $A_1 \otimes A_2$ as parallel execution of A_1 and A_2. Both operations are associative by definition, $\otimes$ is also commutative.

The application algebra can be represented as $\mathcal{H} = [B]$,

where

$$[B] := \bigcup_{i=0}^{\infty} B^i$$

if

$$B := \{B_1, \ldots, B_n\}$$

and :

$$B^0 := \{0\}$$

$$B^1 := B$$

$$B^{i+1} := B^i \otimes B \cup B^i \oplus B \cup B \oplus B^i$$

We shall give B^0 the properties of the *identity* by defining

$$B^0 \oplus A := A \oplus B^0 := B^0 \otimes A := A \quad \text{for all} \quad A \in \mathcal{H},$$

and denote it henceforth by E. B is the union of three mutually exclusive sets B_C, B_D, B_V.

In analogy to the substructure of $\wp$ we can define three subsets of $\mathcal{H}$:

$$\mathcal{H}_V := [B_V]$$

$$\mathcal{H}_D := [B_D \cup B_V] \backslash [B_V]$$

$$\mathcal{H}_C := \mathcal{H} \backslash [B_D \cup B_V]$$

Obviously all three subsets are *subsemigroups* of $\mathcal{H}$, and, in fact, $\mathcal{H}_C$ is an *ideal*.

We shall study the relationship of elements of $\mathcal{H}$ and their homomorphic "values" in the performance algebra $\wp$. In practice the "values" of the generating building blocks can be determined from measurements or deduced from underlying, even more elementary building blocks. In fact, in the section on examples it will become apparent that it is useful to deal with the application algebra (and, similarily, with the implementation algebra) in the context of a decending chain of algebras $\mathcal{H}_1 \supset \mathcal{H}_2 \supset \mathcal{H}_3 \supset \ldots$ where $\mathcal{H}_{i+1}$ is generated from a subset of $\mathcal{H}_i$ and where the values of φ are retained for the generating set of $\mathcal{H}_{i+1}$.

Let $\infty \geq p > q \geq 1$. Then we shall define a *homomorphism* $\varphi : \mathcal{H} \to (\wp, \otimes^q, \otimes^p)$ in the following way:

We define φ on B by assigning a value in $\wp$ to each element of B:

$$\varphi(B_j) := (r_j, w_j, u_j)$$

with : $u_j := \begin{cases} 1 & \text{for } B_j \in B_C \\ 0 & \text{otherwise} \end{cases}$ and $w_j := \begin{cases} i & \text{for } B_j \in B_V \\ \in \mathbb{R}_+ & \text{otherwise} \end{cases}$ where i denotes $\sqrt{-1}$

In other words $\varphi(B_V) \subset V$, $\varphi(B_C) \subset C$, $\varphi(B_D) \subset D$.

If the φ-values of A_1, A_2 are known, the values for their products are:

$$\varphi(A_1 \oplus A_2) := \varphi(A_1) \otimes^q \varphi(A_2)$$

$$\varphi(A_1 \otimes A_2) := \varphi(A_1) \otimes^p \varphi(A_2)$$

We note that $\varphi(E) = e$, where e denotes the identity of $\wp$.

Furthermore $\varphi(\mathcal{H}_V) \subset V$, $\varphi(\mathcal{H}_C) \subset C$, $\varphi(\mathcal{H}_D) \subset D$.

Via φ we can induce a preordering on $\mathcal{H}$ in the following way:

$$A_1 \leq_\varphi A_2 :\Leftrightarrow \varphi(A_1) \leq \varphi(A_2)$$

$$A_1 \lesssim_\varphi A_2 :\Leftrightarrow \varphi(A_1) \lesssim \varphi(A_2)$$

Obviously $(\mathcal{H}, \leq_\varphi)$ becomes a *preordered semigroup* with respect to $\oplus$ and $\otimes$ and the same is true for $(\mathcal{H}, \lesssim_\varphi)$, if $p < \infty$, as from $A_1 \leq A_2$ it follows that

$$\varphi(A \oplus A_1) = \varphi(A) \otimes^q \varphi(A_1) \lesssim \varphi(A) \otimes^q \varphi(A_2) = \varphi(A \oplus A_2)$$

and hence

$$A \oplus A_1 \lesssim_\varphi A \oplus A_2.$$

The operation $\otimes$ and the relation $\leq_\varphi$ can be treated analogously.

Rather than comparing arbitrary algorithms, however, we are more interested in comparing those that in some sense "solve the same problem".

For this purpose we introduce certain *"derivation rules"* that describe replacements of subexpressions in elements of $\mathcal{H}$:

(1) $A_1 \otimes A_2 \rightarrow_0 A_1 \oplus A_2$

(2) $(A_1 \oplus A_2) \otimes (A_3 \oplus A_4) \rightarrow_0 (A_1 \otimes A_3) \oplus (A_2 \otimes A_4)$

where $A_i \in \mathcal{H}$, $i = 1, \dots, 4$ and where $A_1, A_2 \neq E$ in rule (1) and not more than one $A_i = E$ in rule (2).

We shall use these rules in the following context:

If $A \in \mathcal{H}$ contains a subexpression of the left hand side of either rule then this subexpression can be substituted by the expression on the right hand side of the same rule, yielding an element $A' \in \mathcal{H}$. The relationship between A and A' will be denoted by $A \rightarrow_0 A'$ ("A' directly derived from A "). If A' is derived from A by applying a number of rules, we denote this relationship by $A \xrightarrow{*}_0 A'$ (" A' derived from A ").

The process of derivation induces a partial order on $\mathcal{H}$ that is compatible with the semigroup operations:

Theorem 7:

$\mathcal{H}$ is an ordered semigroup w.r.t $\xrightarrow{*}_0$, and from $A \xrightarrow{*}_0 A'$ it follows that $A \geq_\varphi A'$.

∎

If $A \xrightarrow{*}_0 A'$, we can say that A and A' solve, in a sense, the same problem. The derivation graph "below" each $A \in \mathcal{H}$ contains a finite number of nodes.

The following theorem is the "algorithmic" counterpart to the "Outer Loop Parallelization Theorem" in the previous section.

Theorem 8:

Let $A_{ij} \in \mathcal{H}$ and $A := \prod_{j=1}^{n}{}^{(\infty)} \left(\sum_{i=1}^{m}{}^{(1)} A_{ij} \right)$ and $A' := \sum_{i=1}^{m}{}^{(1)} \left(\prod_{j=1}^{n}{}^{(\infty)} A_{ij} \right)$

Then $A \xrightarrow{*}_0 A'$ and $A \geq_\varphi A'$.

∎

Besides rules (1) and (2) there may be additional knowledge about different algorithms (i.e. different elements of $\mathcal{H}$) that solve the same problem. Such a knowledge can be represented by introducing finitely many additional derivation rules, compatible with the semigroup operations, such that left hand and right hand side of these rules contain only elements of the same subsemigroup $\mathscr{H}_c$, $\mathscr{H}_p$, or $\mathscr{H}_v$.

If we denote the union of the "new" with the "old" rules by $\rightarrow$, we obviously have $A \xrightarrow{}_0 A' \Rightarrow A \rightarrow A'$ (thus we call $\rightarrow$ "coarser" than $\rightarrow_0$).

Corollary 2:

Let the generating set B of $\mathscr{H}$ be the union of two mutually exclusive sets B^A and B^S and let a set of derivation rules, coarser than $\rightarrow_0$, be defined in the following way:

Let there be a number K such that for all positive integers $k \le K$ and all $S \in B^S$ there is a $A^k \in B^A$ such that $k \otimes S \rightarrow A^k$.

Finally let $A := \sum_{i=1}^{m} {}^{(1)}S_{i,}$, $S_i \in B^S$ then

$$k \otimes A \xrightarrow{\ } \sum_{i=1}^{m} {}^{(1)}A_i^k$$

■

The previous corollary gives us, at least in principle, the possibility to generate by means of our derivation rules "vectorized" portions of certain algorithms.

Notational remark

The expression $\sum^{m} {}^{(1)}A_i^k$ we shall frequently denote by $(k \otimes A)^T$ alluding to the transpose of a matrix.
■

Remark:
In the subsequent part of this section we shall always assume the homomorphism φ as being defined with $q = 1$ and $p = \infty$, i.e. $\varphi(A_1 \oplus A_2) = \varphi(A_1) \oplus \varphi(A_2)$ and $\varphi(A_1 \otimes A_2) = \varphi(A_1) \otimes \varphi(A_2)$.
■

We shall now turn to the properties of an important class of expressions, the *pipelines*:

Let $A_i \in \mathscr{H}$, $i = 1, \dots, p$ then we define

$$P(n, A_1, \dots, A_p) := \left[\sum_{i=1}^{\min(n,p-1)} {}^{(1)} \left(\prod_{j=1}^{i} {}^{(\infty)} A_j \right) \right]$$

$$\oplus \left[\sum_{i=1}^{|n-p+1|} {}^{(1)} \left(\prod_{j=\max(0,i+\min(n,p-1)-n)}^{\min(p-1,n+i-1)} {}^{(\infty)} A_{j+1} \right) \right]$$

$$\oplus \left[\sum_{i=\max(1,p-n)}^{p-1} {}^{(1)} \left(\prod_{j=i}^{p-1} {}^{(\infty)} A_{j+1} \right) \right]$$

If $n \geq p - 1$ this reduces to the more amiable form

$$P(n,A_1, \dots ,A_p) =$$

$$\left[\sum_{i=1}^{p-1} {}^{(1)}(\prod_{j=1}^{i} {}^{(\infty)}A_j) \right] \oplus \left[\sum_{i=1}^{n-p+1} {}^{(1)}(\prod_{j=0}^{p-1} {}^{(\infty)}A_{j+1}) \right] \oplus \left[\sum_{i=1}^{p-1} {}^{(1)}(\prod_{j=i}^{p-1} {}^{(\infty)}A_{j+1}) \right]$$

For the two limiting cases $p = 1$ and $n = 1$ we have $P(n,A) = n \oplus A$ and

$$P(1,A_1, \dots , A_p) = \sum_{i=1}^{p} {}^{(1)}A_i.$$

We call p the horizontal length, and n the vertical length of the pipeline. If the vertical lengths of two pipelines match, we can define an operation $\copyright$ between them which we shall call **chaining**. in the following way:

$$P(n, A_1, \dots , A_p) \copyright P(n, B_1, \dots , B_q) := P(n, A_1, \dots , A_p, B_1, \dots , B_q)$$

Obviously the operation $\copyright$ is associative on the set of "conformable" pipelines(i.e. those that have the same vertical length), which, of course, is a subset of $\mathscr{H}$.

Let us assume that $\sigma(\varphi(A_i)) = \sigma, u(\varphi(A_i)) = u, i = 1, \dots ,p$. Let $S \in \mathscr{H}_v$ and $\delta := \sigma(\varphi(S))$.

The process of defining Hockney's $n_{1/2}$ can be described in terms of operations within $\wp$ using the following general scheme (which we shall subsequently refer to as "Hockney scheme"):

Let $(g_n)_{n=1}^{\infty}$ be a sequence in $\wp$ where we assume that the sequence $(\frac{1}{n} \oplus g_n)_{n=1}^{\infty}$ is convergent and

$$g_\infty := \lim_{n \to \infty} (\frac{1}{n} \oplus g_n) .$$

Let further $0 < \lambda \leq 1$, then n_λ is determined by the equation:

$$\frac{\lambda}{n_\lambda} \oplus g_{n_\lambda} = \lambda \otimes g_\infty$$

For $g_n := \varphi (S \oplus P(n, A_1, \dots , A_p))$ we obtain:

$$n_{1/2} = \delta/\sigma + p - 1$$

and we finally derive **Hockney's Law** for a pipeline [7]:

$$\sigma(\varphi(S \oplus P(n, A_1, \dots , A_p))) = \frac{w_\infty}{r_\infty} (n_{1/2} + n)$$

If δ/σ is an integer, say q, it is easily verified that

$$\varphi(P(n, S_1, \dots , S_q) \copyright P(n, A_1, \dots , A_p)) = g_n$$

where $P(n, S_1, \dots , S_q) \in \mathscr{H}_v$ and $\sigma(\varphi(S_i)) = \sigma, i = 1, \dots ,q$, i.e. the same timing behaviour is observed if we interpret g_n as being generated from chaining a pipeline of delay elements with $P(n, A_1, \dots , A_q)$.

Hockney's scheme is used subsequently to discuss a number of expressions involving pipelines:

1. Let us look at the behaviour of the chaining of two pipelines

$$P(n, A_1, \ldots, A_p) \,©\, P(n, A_{p+1}, \ldots, A_{p+q}) = P(n, A_1, \ldots, A_{p+q})$$

where

$$u(\varphi(A_i)) = u', \quad i = 1, \ldots, p \text{ and}$$

$$u(\varphi(A_i)) = u'', \quad i = p + 1, \ldots, p + q,$$

such that $u' \vee u'' = 1$ and where $\sigma(\varphi(A_i)) = \sigma, \ i = 1, \ldots, p + q$. Let us use the following notation:

$$\varphi\big(P(n, A_1, \ldots, A_p)\big) = g'_n.$$

$$\varphi\big(P(n, A_{p+1}, \ldots, A_{p+q})\big) = g''_n.$$

$$\varphi\big(P(n, A_1, \ldots, A_{p+q})\big) = g_n.$$

Then we can easily derive a number of identities, even though φ is not a homomorphism with respect to $©$.

$$g_n = \left(n\frac{u'r_\infty' + u''r_\infty''}{n + p + q - 1}, \ n(u'w_\infty' + u''w_\infty''), \ 1 \right)$$

$$r_\infty = u'r'_\infty + u''r'_\infty$$

$$n_{1/2} = n'_{1/2} + n''_{1/2} + 1$$

2. Another expression that is of some importance in practice is the following

$P(n, A_1, \ldots, A_p) \oplus P(n, A_{p+1}, \ldots, A_{p+q})$ which is not a pipeline itself but can be treated by the general Hockney scheme we described above. We obtain:

$$g_\infty = g'_\infty \oplus g''_\infty \text{ and}$$

$$n_{1/2}\sigma(g_\infty) = n'_{1/2}\sigma(g'_\infty) + n''_{1/2}\sigma(g''_\infty)$$

3. For the construct $P(n, A_1, \ldots, A_p) \otimes P(n, A_{p+1}, \ldots, A_{p+q})$ we obtain

$$g_\infty = g'_\infty \otimes g''_\infty \text{ but no convenient expression for } n_{1/2} \text{ is available.}$$

4. Let us finally look at: $g_n^N := N \otimes g_n$ where $g_n := \varphi\big(P(n, A_1, \ldots, A_p)\big)$.

We obviously have $g_\infty^N = N \otimes g_\infty$, $r_\infty^N = N \cdot r_\infty$, and $n_{1/2}^N = n_{1/2}$.

However, we can also look at the Hockney scheme from another angle by using a different interpretation, setting $g_{N \cdot n}' := g_n^N$. The result is $r_\infty^* = N \cdot r_\infty$ but also $n_{1/2}^* = N \cdot n_{1/2}$

The latter interpretation is frequently applied if multiple pipelines are used for the same operation type in vector machines.

Remark:
In order to describe the arithmetic pipelines of vector machines we can set $w_\infty = 1$ and interpret σ as the cycle time.

Let us now turn to the question of providing a "sufficient" amount of granularity in order to compensate for the **dispatching overhead** in a multitasking environment. This question has been studied by Hockney [8].

Let

$$S \in \mathcal{H}_V \text{ and } A^n := S \oplus \prod_{i=1}^{p}{}^{(\infty)}(n \oplus A_i)$$

Let further $\varphi(A^n) =: g_n$ and $\sigma(\varphi(S)) =: \delta$ and $\bar{g} := \frac{1}{p} \otimes \prod_{i=1}^{p}{}^{(\infty)} \varphi(A_i)$. If we apply Hockney's scheme we obtain:

$$g_\infty = p \otimes \bar{g} \text{ and } n_{1/2} = \delta / \bar{\sigma}$$

The quantity $n_{1/2} w_\infty = r_\infty \delta$ describes the amount of useful work that could have been done during the delay δ, and is identical with **Hockney's** $s_{1/2}$.

If we set $s(n) := n w_\infty$ we obtain: $\sigma(g_n) = \frac{1}{r_\infty}(s_{1/2} + s(n))$

Remark:
The dispatching delay can often be modelled by $S = S_0 \oplus (p \oplus S_T)$, where there is a fixed cost S_0 and a cost S_T for each dispatched task.

■

4. The Implementation Algebra

Unfortunately not all the parallelism, denoted by $\otimes$ in the previous section, evident in an algorithm, can be simultaneously executed in a real machine. In this respect the homomorphism φ in its unlimited use as described above represents the situation of the ideal machine, i.e. with infinite parallelism.

In the following discussion we shall attempt to describe a more "realistic" situation in algebraic terms.

Let $\mathcal{R}$ (the "implementation algebra") be a finitely generated free semigroup with respect to $\oplus$ and $\otimes$ such that there exists a **weak homomorphism** $\psi_\mathcal{R} : \mathcal{H} \to P(\mathcal{R})$, where $P(\mathcal{R})$ denotes the power set of $\mathcal{R}$, with the following properties:

(1a) $\quad \psi_\mathcal{R}(A_1) \oplus \psi_\mathcal{R}(A_2) \subset \psi_\mathcal{R}(A_1 \oplus A_2)$

(1b) $\quad \psi_\mathcal{R}(A_1) \otimes \psi_\mathcal{R}(A_2) \subset \psi_\mathcal{R}(A_1 \otimes A_2)$

(2) $\quad \psi_\mathcal{R}(\mathcal{H}_C) \subset \mathcal{R}_C, \psi_\mathcal{R}(\mathcal{H}_D) \subset \mathcal{R}_D, \psi_\mathcal{R}(\mathcal{H}_V) \subset \mathcal{R}_V$

$P(\mathcal{R})$ can be made into a semigroup with respect to $\oplus$ and $\otimes$ in the usual way, i.e.: let $R_1, R_2 \in P(\mathcal{R})$ then $R_1 \oplus R_2 := \{M_1 \oplus M_2 | M_i \in R_i, i = 1, 2\}$. The operation $\otimes$ can be treated in an analogous way.

Let us assume that a homomorphism $\varphi_\mathcal{R} : \mathcal{R} \to \wp$ has been defined in the spirit of the previous section, and that via $\varphi_\mathcal{R}$ preorderings $\leq_{\varphi_\mathcal{R}}$ and $\leq_{\varphi_\mathcal{R}}$ have been introduced on $\mathcal{R}$. These preorderings can easily be "lifted" to $P(\mathcal{R})$ in the following way:

Let $R_1, R_2 \in P(\mathcal{R})$ then $R_1 \leq_{\varphi_{\mathcal{R}}} R_2$ if there is a $M_2 \in R_2$ such that $M_1 \leq_{\varphi_{\mathcal{R}}} M_2$ for all $M_1 \in R_1$. For convenience we shall subsequently only use $\leq_{\varphi_{\mathcal{R}}}$, but $\leq_{\varphi_{\mathcal{R}}}$ with its known limitations could have been treated in an analogous manner.

We still need another restriction on the behaviour of $\psi_{\mathcal{R}}$:

(3) $\psi_{\mathcal{R}}(A_1) \leq_{\varphi_{\mathcal{R}}} \psi_{\mathcal{R}}(A_2)$ for $A_1, A_2 \in \mathcal{H}$ then $\psi_{\mathcal{R}}(A_1 \otimes A) \leq_{\varphi_{\mathcal{R}}} \psi_{\mathcal{R}}(A_2 \otimes A)$

    for all $A \in \mathcal{H}$

where $\psi_{\mathcal{R}}$ has an analogous behaviour with respect to the operator $\oplus$.

Remark:
If $\psi_{\mathcal{R}}$ is a (set valued) homomorphism then property (3) comes for free.
∎

We are now in the position to define a preordering on $\mathcal{H}$ with respect to $\psi_{\mathcal{R}}$.

Definition:

Let $A_1, A_2 \in \mathcal{H}$ then $A_1 \leq_{\psi_{\mathcal{R}}} A_2 :\Leftrightarrow \psi_{\mathcal{R}}(A_1) \leq_{\varphi_{\mathcal{R}}} \psi_{\mathcal{R}}(A_2)$
∎

Clearly, because of (3), $(\mathcal{H}, \leq_{\psi_{\mathcal{R}}})$ is a preordered semigroup.

Unfortunately we have no constructive way of desrcibing $\psi_{\mathcal{R}}$. What we can do however, is giving a constructive description of a *lower bound* $\hat{\psi}_{\mathcal{R}}$ of $\psi_{\mathcal{R}}$ in the following way, if the values of $\psi_{\mathcal{R}}$ are known on the generating set of $\mathcal{H}$:

Let $\hat{\psi}_{\mathcal{R}}(B_i) := \max \psi_{\mathcal{R}}(B_i), i = 1, \dots, m$, where the maximum is taken with respect to $\leq_{\varphi_{\mathcal{R}}}$. $\hat{\psi}_{\mathcal{R}}$ can then be extended to all of $\mathcal{H}$ by homomorphic continuation.

Theorem 9:

Let $A \in \mathcal{H}$ then $\hat{\psi}_{\mathcal{R}}(A) \in \psi_{\mathcal{R}}(A)$ and $\hat{\psi}_{\mathcal{R}}(A) \leq_{\varphi_{\mathcal{R}}} \psi_{\mathcal{R}}(A)$.
∎

$\hat{\psi}_{\mathcal{R}}$ is, loosely speaking, the best pessimistic homomorphic selection for $\psi_{\mathcal{R}}$.

Obviously we also have for the composition $\hat{\varphi}_{\mathcal{R}} := \varphi_{\mathcal{R}} \circ \hat{\psi}_{\mathcal{R}}$ the inequality: $\hat{\varphi}_{\mathcal{R}}(A) \leq \varphi_{\mathcal{R}}(\psi_{\mathcal{R}}(A))$ for all $A \in \mathcal{H}$, if $\leq$ is lifted to the powerset of $\wp$ in a way we have described above for $\mathcal{R}$. Furthermore $\hat{\varphi}_{\mathcal{R}}$ has the properties of φ in the previous section.

Let $J \subset \mathcal{R}$ be the subset of *admissible elements*, those that can be "effectively" executed in the sense that the parallelism displayed in the expressions contained in J does not exceed the resources of a machine that we associate with $\mathcal{R}$. If we just worry about finite parallelism, we may make the assumption that J is a semigroup w.r.t. $\oplus$ (if other resources like memory are taken into consideration, this may not be a viable assumption.)

We can now define the admissible subset of $\mathcal{H}$ with respect to $\psi_{\mathcal{R}}$ by:

$$J_{\psi_{\mathscr{R}}} := \psi_{\mathscr{R}}^{-1}(J \cap \psi_{\mathscr{R}}(\mathscr{K}))$$

It is convenient to assume a "regular" behaviour for $\psi_{\mathscr{R}}$ with respect to J:

(4) $\psi_{\mathscr{R}}(A) \cap J \neq \emptyset \Rightarrow \psi_{\mathscr{R}}(A) \subset J$

In other words, if A has an admissible implementation, then all implementations of A are admissible.

With property (4) in mind we can conveniently describe a process that is frequently performed during **benchmarking**, where within a given class of algorithms that solve the same problem the one most suited for a given machine is selected. In the framework of our algebraic apparatus we can formulate this problem as finding:

$$A \in \max(K \cap J_{\psi_{\mathscr{R}}}) := \{A \in K \cap J_{\psi_{\mathscr{R}}} \mid K \cap J_{\psi_{\mathscr{R}}} \lesssim_{\psi_{\mathscr{R}}} A\}$$

where K is a given class generated by some $\rightarrow$ coarser than $\rightarrow_0$. If this problem seems too hard, we can at least try an "optimization level 0"-approximation and search for a member of the set

$$\{A \in K \cap J_{\psi_{\mathscr{R}}} \mid \hat{\psi}_{\mathscr{R}}(K \cap J_{\psi_{\mathscr{R}}}) \lesssim_{\varphi_{\mathscr{R}}} \hat{\psi}_{\mathscr{R}}(A)\}$$

where (4) together with Theorem 9 ensures that $\hat{\psi}_{\mathscr{R}}(A) \in J$.

Of course, if the behaviour on different machines (different $\mathscr{R}$'s and different $\psi_{\mathscr{R}}$'s) is compared, it is important to keep K fixed.

5. Examples

In the first example, a (memory to memory) array operation is generated from vector instructions whose performance values in turn are obtained from the considerations on pipelines in Section 3.

In the second example we discuss a matrix multiplication where two different paths are pursued: in the first case the matrix multiply is generated directly from vector instructions, whereas in the second case it is built up out of (memory to memory) array operations. This second, "context-free" approach does not take advantage of the fact that subsequently needed data have already been moved to a higher level of the memory hierarchy, and is thus less efficient.

In the third example we deal with a macro-pipelined Gauß-Seidel Iteration for Laplaces equation on a rectangle. We investigate the implementation on two different types of architectures a local-memory multiprocessor and a shared memory vector machine.

Example 1: "Algebraic Bar Diagrams"

We shall describe a number of implementation patterns on various vector machines for the array operation **Vector plus Scalar times Vector**

Traditionally, bar diagrams have been in use to estimate the approximate number of machine cycles per floating operation by displaying the type of overlap between elementary vector instructions:

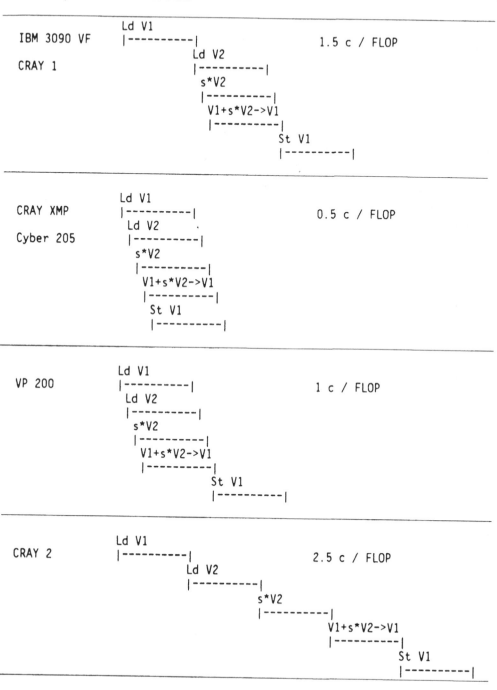

```
                    Ld V1
IBM 3090 VF         |----------|                            1.5 c / FLOP
                           Ld V2
CRAY 1                     |----------|
                             s*V2
                             |----------|
                               V1+s*V2->V1
                               |----------|
                                       St V1
                                       |----------|
```

```
                    Ld V1
CRAY XMP            |----------|                            0.5 c / FLOP
                    Ld V2
Cyber 205           |----------|
                      s*V2
                      |----------|
                      V1+s*V2->V1
                      |----------|
                        St V1
                        |----------|
```

```
                    Ld V1
VP 200              |----------|                            1 c / FLOP
                    Ld V2
                    |----------|
                      s*V2
                      |----------|
                      V1+s*V2->V1
                      |----------|
                              St V1
                              |----------|
```

```
                    Ld V1
CRAY 2              |----------|                            2.5 c / FLOP
                          Ld V2
                          |----------|
                                s*V2
                                |----------|
                                    V1+s*V2->V1
                                    |----------|
                                            St V1
                                            |----------|
```

Let $n = m \cdot lv + k$, where lv denoted the length of the vector register (or maximum pagesize on the CYBER 205) and $1 \leq k \leq lv$.

For simplicity of notation we use the same algebra $\mathcal{R}$ for all machines we discuss below, where actually we should have assigned to each of them the appropriate subset with machine specific φ-values.

We assume that vector instructions for arithmetic and data movement operations are executed by pipelines. Their φ-values are defined in accordance with our results on pipelines in section 3. For $S_j \in V$ we have

$$\varphi(S_j) := (\frac{1}{\delta_j}, i, 0), \quad j = 0,1$$

It is important to note that the chaining in the expressions we deal with in this section takes place in the algebra of "micro operations" that constitute the vector operations whereas in the algebra of vector instructions these chainings cannot be expressed in terms of $\oplus$ and $\otimes$ and thus play a merely mnemonic role.

The general form for the various implementations is:

$$S_0 \oplus \sum_{i=1}^{m+1} {}^{(1)}S_1 \oplus M(n_i, \alpha, V_1^i, V_2^i)$$

where $n_i = lv, i = 1, \dots, m$ and $n_{m+1} = k, \ 1 \leq k \leq lv$ and where

$$\varphi\left(M(k, \alpha, V_1^i, V_2^i)\right) := \left(\frac{r_\infty^M k}{k + n_{1/2}^M}, \ w_\infty^M k, \ 1\right)$$

Type 1 (CRAY 1, IBM 3090 VF)

$$M(k, \alpha, V_1^i, V_2^i) :=$$

$$LD(k, V_1^i) \oplus (LD(k, V_2^i) \copyright MPY(k, \alpha, V_2^i, V) \copyright ADD(k, V, V_1^i, V_1^i)) \oplus STR(k, V_1^i)$$

Then

$$n_{1/2}^M = \frac{1}{3} \cdot (2p_l + p_m + p_a + p_s - 3)$$

$$r_\infty^M = \frac{2}{3} r_\infty$$

Type 2 (CRAY-XMP, CYBER 205)

$$M(k, \alpha, V_1^i, V_2^i) :=$$

$$LD(k, V_1^i) \otimes (LD(k, V_2^i) \copyright MPY(k, \alpha, V_2^i, V) \copyright ADD(k, V, V_1^i, V_1^i) \copyright STR(k, V_1^i))$$

Then

$$n_{1/2}^M = p_l + p_m + p_a + p_s - 1$$

$$r_\infty^M = 2r_\infty$$

Type 3 (Fujitsu VP 200)

$$M(k, \alpha, V_1^i, V_2^i) := $$

$$LD(k, V_1^i) \otimes (LD(k, V_2^i) \copyright MPY(k, \alpha, V_2^i, V) \copyright ADD(k, V, V_1^i, V_1^i)) \oplus STR(k, V_1^i)$$

Then

$$n_{1/2}^M = \frac{1}{2} \cdot (p_l + p_m + p_a + p_s - 2)$$

$$r_\infty^M = r_\infty$$

Type 4 (CRAY 2)

$$M(k, \alpha, V_1^i, V_2^i) := $$

$$LD(k, V_1^i) \oplus LD(k, V_2^i) \oplus MPY(k, \alpha, V_2^i, V) \oplus ADD(k, V, V_1^i, V_1^i) \oplus STR(k, V_1^i)$$

Then

$$n_{1/2}^M = \frac{1}{5} \cdot (2p_l + p_m + p_a + p_s - 5)$$

$$r_\infty^M = \frac{2}{5} r_\infty$$

For the timing of a fairly general array operation OP we obtain:

$$\sigma(\varphi(OP(n))) = \delta_0 + \sum_{i=1}^m \left(\delta_1 + \frac{w_\infty^{OP}}{r_\infty^{OP}}(lv + n_{1/2}^{OP}) \right) + \delta_1 + \frac{w_\infty^{OP}}{r_\infty^{OP}}(k + n_{1/2}^{OP})$$

Obviously this is nothing but the familiar "sawtooth"-function (see e.g. [2]).

$$t(n) := a + b \cdot n + d \cdot [n/lv]$$

where $[x]$ denotes the smallest integer greater than or equal to x.

By comparison we obtain $a = \delta_0$, $b = \dfrac{w_\infty^{OP}}{r_\infty^{OP}}$, and $d = \delta_1 + \dfrac{w_\infty^{OP}}{r_\infty^{OP}} n_{1/2}^{OP}$.

If we set $g_n := \varphi(OP(n))$ we can obtain the effective rate from Hockney's process:

$$r_\infty^{OP,eff} = \frac{r_\infty^{OP}}{\left(1 + \dfrac{\delta_1/\sigma_\infty^{OP} + n_{1/2}^{OP}}{lv} \right)}$$

$1/r_\infty^{OP,eff}$ also determines the slope of a linear approximation to the sawtooth-function in the least squares sense.

Example 2: Matrix Multiply

Let C be a $(l \cdot n)$, A a $(l \cdot m)$, and B a $(m \cdot n)$ Matrix. We are looking at columnwise implementations of a matrix multiplication $C = A \cdot B$, i.e.

$$C_j = \sum_{k=1}^{m} B_{kj} A_k$$

where the columns C_j of C are written as a linear combination of the columns A_k of A.

If we ignore the initialization of C a corresponding element of $\mathcal{H}$ would take the following form:

$$MMPY(C,A,B,l,m,n) := \sum_{j=1}^{n}{}^{(1)} \sum_{k=1}^{m}{}^{(1)} SAXPY(l,B_{kj}, A_k,C_j,C_j)$$

Let us assume that we are dealing with a cache based system like the IBM 3090, and that the number k_{eff} is such that $k_{eff} \cdot lv$ does not exeed the effective cache size. For ease of notation we further assume $l = \lambda \cdot lv$ and $m = k_{eff} \cdot \mu$.

FORTRAN aspects and the various implications of memory hierarchies for this example are discussed in [12].

A piece of FORTRAN code that takes advantage of the ability of the IBM compiler to vectorize the "outer loop" looks like this:

```
      MUE = M/KEFF
      DO 400 I = 1,L
      DO 300 IR = 1,MUE
      DO 200 J = 1,L
      S = 0.D0
      DO 100 K = IR * KEFF + 1 , (IR + 1) * KEFF
      S = S + B(K,J) * A(I,K)
100   CONTINUE
      C(I,J) = C(I,J) + S
200   CONTINUE
300   CONTINUE
400   CONTINUE
```

In the subsequent "implementation" we will only take reuse of data in vector registers and cache into account, and we neglect the effect of loading the scalar factors from matrix B:

$$MMPY :=$$

$$\sum_{i=1}^{\lambda}{}^{(1)} \sum_{r=0}^{\mu-1}{}^{(1)} \sum_{j=1}^{n}{}^{(1)} \left\{ LD(lv,C_j^i) \oplus MMPC(C_j^i,A,B,i,r,j,lv,k_{eff}) \oplus STR(lv,C_j^i) \right\}$$

where for $j = 1$

$$MMPC(C_j^i,A,B,i,r,j,lv,k_{eff}) :=$$

$$\sum_{k=r\cdot k_{eff}+1}^{(r+1)\cdot k_{eff}} {}^{(1)}\Big(LD(Iv, A_k^i)\copyright MPY\copyright ADD(Iv,B_{kj}, A_k^i, C_j^i, C_j^i)\Big)$$

describes the initial loading of the cache with the (i, r)-th "window" of A chained with the linear combination of the corresponding column segments of A, where the result is written into a vector register, and for $j = 2,....n$

$$MMPC(C_j^i,A,B,i,r,j,Iv,k_{eff}) := \sum_{k=r\cdot k_{eff}+1}^{(r+1)\cdot k_{eff}} {}^{(1)}\Big(MPY\copyright ADD(Iv,B_{kj}, A_k^i, C_j^i, C_j^i)\Big)$$

stands for the linear combination of the column segments of A that are already contained in the cache.

It is straight forward to obtain

$\varphi(MMPY) =$

$$\left(\frac{2r_\infty}{\frac{1}{Iv}\left(\frac{Iv+p_I-1}{k_{eff}} + \frac{p_I-1}{n}\right) + \frac{Iv+p_a+p_m-1}{Iv} + \frac{1}{k_{eff}}\cdot\frac{Iv+p_s-1}{Iv}} , 2Imn , 1\right)$$

An "optimization level 0 - implementation", in the sense this term is used in the previous section, will look like this:

$$MMPY := \sum_{j=1}^{n} {}^{(1)}\sum_{k=1}^{m} {}^{(1)} MAD(I,B_{kj}, A_k,C_j,C_j)$$

where

$$MAD(Iv,B_{kj},A_k,C_j,C_j) :=$$

$$\sum_{i=1}^{\mu} {}^{(1)} LD(Iv,C_j^i) \oplus \Big(LD(Iv,A_k^i)\copyright MPY\copyright ADD(Iv,B_{kj},C_j^i,C_j^i)\Big)\oplus STR(Iv,C_j^i)$$

If we make use of the discussion of algebraic bar diagrams we obtain:

$$\varphi(MMPY) = \left(\frac{\frac{2}{3}r_\infty Iv}{\frac{1}{3}(2p_I+p_a+p_m+p_s-3)+Iv} , 2Imn , 1\right)$$

Thus the first implementation is approximately a factor of 3 better than the second implementation.

Example 3

Let us consider Laplaces Equation on a rectangle with Dirichlet boundary conditions. Discretization in the usual way with equal spacing in both directions yields the following well known discretized equation:

$$\frac{1}{h^2}\left(- 4x_{ij} + x_{i-1,j} + x_{i+1,j} + x_{i,j+1} + x_{i,j-1}\right) = 0$$

where boundary values for the $N \cdot M$ grid are given.

Iterative methods are among the most popular to solve the this sparse linear system of equations, and among those the Gauß-Seidel iteration with its various variants has been frequently in use:

$$x_{ij}^{n+1} = \frac{1}{4}\left(x_{i-1,j}^{n+1} + x_{i+1,j}^{n} + x_{i,j+1}^{n} + x_{i,j-1}^{n+1}\right)$$

Looking at one iteration step at a time, the Gauß-Seidel iteration is, because of its obvious dependences, not parallelizable (or vectorizable), if one progresses in a rowwise or columnwise fashion. Marching in a diagonal direction or adopting a checkerboard- numbering scheme are well known remedies (even though the latter of the two is, strictly speaking, a modification of the algorithm).

We shall demonstrate a "macro-pipeline" approach that goes across several iteration steps, but implies no change of the original Gauß-Seidel Iteration.

Let A, be described by the following piece of FORTRAN-code:

```
      SUBROUTINE GSL(I ...
      DO 10 J = 2,M-1
      X(I,J) = .25 * ( X(I-1,J) + X(I,J-1) + X(I,J+1) + X(I+1,J) )
   10 CONTINUE
      RETURN
      END
```

If we adopt a zebra-pattern in a rowwise manner we end up with the following parallel-sequential expression for $N/2 + L$ iterations (N assumed even):

$$\sum_{k-1}^{N/2-1}{}^{(1)}\left[\left(\prod_{i-1}^{k}{}^{(\infty)}A_{2i}\right)\oplus\left(\prod_{i-1}^{k}{}^{(\infty)}A_{2i+1}\right)\right]$$

$$\oplus\sum_{k-1}^{L}{}^{(1)}\left[\left(\prod_{i-1}^{N/2-1}{}^{(\infty)}A_{2i}\right)\oplus\left(\prod_{i-1}^{N/2-1}{}^{(\infty)}A_{2i+1}\right)\right]$$

$$\oplus\sum_{k-2}^{N/2-1}{}^{(1)}\left[\left(\prod_{i-k}^{N/2-1}{}^{(\infty)}A_{2i}\right)\oplus\left(\prod_{i-k}^{N/2-1}{}^{(\infty)}A_{2i+1}\right)\right]$$

The strong resemblance of this expression to pipelines discussed in section 3 is obvious.

Let us now take a look at various implementations.

First we assume a multiprocessor with local memories and the ability to communicate with the nearest left and right neighbour.

If $N/2 - 1$ processors are available we can assign tasks A_{2i} and A_{2i+1} to the i-th processor P_i, where we require that rows $X_{2i-1}, \ldots, X_{2i+2}$ are contained in its local memory.

Results are written to the left (A_{2i}) or right (A_{2i+1}) neighbour as soon as they are available:

$$A_{2i} := GSL(2i) \textcircled{C} WRT(X_{2i} \text{ to } P_{i-1})$$

$$A_{2i+1} := GSL(2i + 1) \textcircled{C} WRT(X_{2i+1} \text{ to } P_{i+1})$$

At this degree of resolution we do not model data movements within a single processor.

If only v processors, where $v \cdot K = N - 2$, K even, are available, an implementation for

$$\prod_{\mu = (i-1) \cdot K/2}^{i \cdot K/2} {}^{(\infty)}A_{2\mu} \text{ on } P_i$$

could be

$$\left\{ GSL((i - 1) \cdot K + 2) \textcircled{C} WRT(X_{(i-1) \cdot K + 2} \text{ to } P_{i-1}) \right\} \oplus \sum_{\mu = (i-1) \cdot K/2 + 2}^{i \cdot K/2} {}^{(1)}GSL(2\mu)$$

Accordingly, for an implementation of

$$\prod_{\mu = (i-1) \cdot K/2}^{i \cdot K/2} {}^{(\infty)}A_{2\mu+1} \text{ on } P_i$$

we can write

$$\left\{ GSL(i \cdot K + 1) \textcircled{C} WRT(X_{i \cdot K + 1} \text{ to } P_{i+1}) \right\} \oplus \sum_{\mu = (i-1) \cdot K/2 + 1}^{i \cdot K/2 - 1} {}^{(1)}GSL(2\mu + 1)$$

Here we have assumed that $X_{(i-1) \cdot K + 1}, \ldots, X_{i \cdot K + 2}$ are contained in the local memory of P_i.

Unfortunately load balancing and datamovements are somewhat more awkward and involved for the head- and tail-phase in the v-processor case.

A second type of architecture we would like to investigate for our problem is a shared-memory vector computer. The discussion in section 3 shows that our macro-pipeline can be written in a "vectorized" manner:

$$\sum_{k=1}^{N/2-1}(1)\left[\left(\prod_{i=1}^{k}{}^{(\infty)}A_{2i}\right)^{\mathsf{T}}\oplus\left(\prod_{i=1}^{k}{}^{(\infty)}A_{2i+1}\right)^{\mathsf{T}}\right]$$

$$\oplus\sum_{k=1}^{L}(1)\left[\left(\prod_{i=1}^{N/2-1}{}^{(\infty)}A_{2i}\right)^{\mathsf{T}}\oplus\left(\prod_{i=1}^{N/2-1}{}^{(\infty)}A_{2i+1}\right)^{\mathsf{T}}\right]$$

$$\oplus\sum_{k=2}^{N/2-1}(1)\left[\left(\prod_{i=k}^{N/2-1}{}^{(\infty)}A_{2i}\right)^{\mathsf{T}}\oplus\left(\prod_{i=k}^{N/2-1}{}^{(\infty)}A_{2i+1}\right)^{\mathsf{T}}\right]$$

The operands of the vector instructions are column vectors with a spacing of 2. For $L \geq N/2$ the average vector length is $\geq N/3$.

6. References

[1]   G. Amdahl
      The Validity of the Single Processor Approach to Achieving Large Scale Comput-
      ing Capabilities
      AFIPS Conference Proceedings, Vol 30, 1967

[2]   I. Bucher
      The Computational Speed of Supercomputers
      Proceedings of the ACM Sigmetrics Conference on Measurements and Modeling
      of Computer Systems (1983) 151-165

[3]   K. Ekanadham
      private communication

[4]   G.E. Forsythe, W.E. Wasow
      Finite-Difference Methods for Partial Differential Equations
      Wiley & Sons, New York, London, Sidney, 1967

[5]   L. Fuchs
      Partially Ordered Algebraic Systems
      Pergamon Press, Oxford, London, New York, Paris, 1963.

[6]   M.D. Hebden
      A Bound on the Difference between Chebyshev Norm and Hölder Norms of a
      Function
      SIAM Journal of Numerical Analysis Vol 8 (1971)

[7]   R.W. Hockney, C.R. Jesshope
      Parallel Computers
      Adam Hilger Ltd, Bristol, 1981

[8]   R.W. Hockney
      $(r_\infty, n_{1/2}, s_{1/2})$ measurements on the 2-CPU CRAY XMP
      Parallel Computing 2 (1985) 1-14

[9]   R.B. Holmes
      A Course on Optimization and Best Approximation
      Lecture Notes in Mathematics 257, Springer Verlag, Berlin, Heidelberg, New York,
      1972

[10]   C.C. Hsiung, W. Butscher
       A Numerical Seismic 3-D-Migration Model for Vector Multiprocessors
       Parallel Computing 1 (1984) 113-120

[11]   D. Kuck
       The Structure of Computers and Computations, Vol 1
       Wiley & Sons, 1978

[12]   B. Liu, N. Strother
       Peak Vector Performance from VS Fortran
       to appear

[13]   E.S. Ljapin
       Semigroups
       American Mathematical Society, Providence, Rhode Island, 1963

[14]   D.G. Luenberger
       Optimization by Vector Space Methods
       Wiley & Sons,1969

[15]   J. Martin, D. Müller-Wichards
       Supercomputer Performance Evaluation: Status and Directions
       Journal of Supercomputing to appear in Vol 1, (1986)

[16]   D. Müller-Wichards
       Performance Estimates for Applications : an Algebraic Framework
       to appear in Parallel Computing

[17]   C.H. Papadimitriou, K. Steiglitz
       Combinatorial Optimization: Algorithms and Complexity
       Prentice Hall, Englewood Cliffs, N.J., 1982

[18]   J. Worlton
       Understanding Supercomputer Benchmarks
       Datamation, Sept 1, (1984)

[19]   A. Wouk
       A Course of Applied Functional Analysis
       Wiley & Sons, 1979